SHOTGUNS

A COMPREHENSIVE GUIDE

by

STEVE MARKWITH

Part of the *Survival Guns* series of books
published by:

Prepper Press
Your Survival Library

PrepperPress.com/SurvivalGuns

ISBN 13: 978-1-939473-21-9

Printed in the United States of America.

Prepper Press Paperback Edition: July 2014

Prepper Press is a division of Kennebec Publishing, LLC

Special thanks to Pyramyd Air for their cooperation in lending photographs to help complete this book.

Disclaimer. This book is intended to offer general guidance relating to firearms. It is sold with the understanding that every effort was made to provide accurate information; however, errors are still possible. The author and publisher make no warrantees or claims as to the truth or validity of the information. The author and publisher shall have neither liability nor responsibility to any person or entity over any loss or damage caused, or alleged to have been caused, directly or indirectly, by the information contained within this book.

ABOUT THE AUTHOR

Steve has a life-long interest in just about all things that shoot, including rifles, shotguns, revolvers, pistols, airguns, and black-powder guns, as well as vertical and horizontal bows.

He began formal firearms training at age eleven during NRA-sanctioned small-bore target rifle events, and was an active hunter by the age of twelve. He began reloading shotgun shells at fourteen, using a hand-held Lee-Loader to feed his addiction.

After joining the U.S. Army, he served two com-

Steve & Beretta

bat tours in Vietnam, gaining experience on numerous military firearms systems during Air Cavalry helicopter operations and ground-based reconnaissance missions.

Upon return to civilian life Steve resumed shooting, participating in NRA bulls-eye, combat-pistol, and trap events. These activities expanded his reloading experience to metallic ammunition and bullet casting. Steve eventually became an NRA-certified Pistol, Rifle and shotgun instructor, as well as a certifying official for state firearms permit applicants. He also worked for a well-known gunsmith and PO Ackley disciple – until an untimely death forced a career change.

Joining a major state correctional agency, Steve was soon appointed as a firearms instructor, eventually assuming control of all state correctional firearms operations. He's still working and holds a master-instructor rating, plus numerous other federal, state, and industry certifications. He has 25-plus years of full-time firearms training experience, and many industry connections.

Steve also has extensive hunting experience in the Northeast, and at other locations throughout the United States. He holds an archery deer record, and actively remains afield on a year-round basis, whether chasing spring turkeys or winter coyotes with night-vision equipped AR-15s. He also writes when time permits, and has had numerous articles published about firearms and the great outdoors.

TABLE OF CONTENTS

INTRODUCTION

Shotguns: A Comprehensive Guide is an extension of our first book, *Survival Guns: A Beginner's Guide.*, In *Survival* Guns, we laid out the groundwork for a practical collection of firearms geared toward "survival" use, chosen from eight key requirements:

1. Whatever we're looking at must be in widespread use.

2. It's got to be something with a solid reputation for dependability.

3. It must be easy to operate.

4. Parts must be readily available.

5. Ammunition must be widely available.

6. It must be easy to maintain.

7. It should accommodate practical accessories.

8. It must represent good value.

In *Survival Guns* we procured a gun safe and framed a small but practical inventory of firearms based on these requirements, adding a shotgun, rimfire rifle, centerfire rifle, and a handgun. The idea was to choose, when possible, types with similar function. Use of each would thus promote skills with another to improve overall proficiency, a seamless transition between firearms. Toward that goal, our initial selections were based on the K.I.S.S. principle, keep it simple stupid. The first gun to go in the safe was a shotgun. What follows is an in-depth examination of this highly versatile and somewhat misunderstood system.

Although it's not as "tacticool" as a tricked out black rifle, a shotgun is one darned useful piece of equipment. A correctly chosen scattergun will digest a huge assortment of ammunition and is capable of taking everything from doves to moose. The business end of a 12-gauge shotgun has a very large hole, which if pointed in the direction of a prospective assailant, is likely to command attention. Many types of ammunition remain scarce, but shotgun shells, particularly in common persuasions like 12-gauge, continue to be available. When not employed to harvest game, it'll serve quite nicely to defend the homestead. We can fire a swarm of multiple pellets with each trigger pull or launch a solid slug, plus all sorts of specialty loads like bean bags, chemical agents, illumination, lock busters, and distraction rounds, just to name a few. If forced to travel light, we may be limited to just one shoulder-fired gun, so given its versatility, a shotgun is a logical first choice.

One common misconception is that you can just point a shotgun in the general direction of a target and make a hit. Not true! A shotgun is a smoothbore firearm, meaning the barrel doesn't have the spiral lands and grooves needed to stabilize a single projectile (an exception being purpose-built slug guns). You'll sometimes hear it called a "scattergun" in reference to its multiple-projectile payload. The pellets emerge in an expanding pattern which, at maximum ranges of 40-50 yards, may saturate several square feet. However, at room-length distances, a pattern may cover just several inches. Patterns are based on a number of factors ranging from gauges to chokes. We can also utilize single projectiles to extend our range or tackle big game. Once understood, the shotgun affords some truly useful options.

Two often cited limitations of a shotgun involve range and ammo capacity.

Let's look at the range issue first. Granted, a shotgun is primarily a short-range weapon; however, a good slug matched to the right barrel will provide useful accuracy toward and even beyond one hundred yards. Some of the modern shot loads produce good patterns at surprising distances, too.

Concerning capacity, even a plain, tubular magazine pump gun is a formidable weapon in the hands of a trained user. With spare shells positioned correctly, an operator can shoot and reload while maintaining a firing position. As long as a shell supply is readily accessible, the gun can be in action.

I'd like to spend some extra time exploring this firearm, mainly because so many folks don't seem to have a firm understanding on things like chokes, shot sizes, shell markings, and gauges. Such information may prove useful when narrowing down a final shotgun choice. It may also come in handy when tracking down ammunition.

As discussed in *Survival Guns, A Beginner's Guide*, whatever we're looking at should be in widespread use. This generally rules out any recent introductions. Well-established firearms have plenty of history behind them and will hold few surprises. Widespread use is an indicator of numerous desirable traits, and pretty much sets the stage for what follows.

CHAPTER 1

SELECTION CRITERIA

Let's expand upon the list of eight key requirements set forth in the Introduction. This criteria follows a list developed in the first edition, and will help steer us towards practical selections.

Whatever we're looking at must be in widespread use. This generally rules out any recent introductions. Well-established firearms have plenty of history behind them and will hold few surprises. Widespread use is an indicator of desirable traits, and sets the stage for the next requirements.

Whatever we choose should be something with a solid reputation for dependability. There are teething pains with many new products. Firearms are no exception. My Firearms Training Unit (FTU) often procures test and evaluation (T&E) samples which get a thorough work through from multiple users. It's not uncommon to run into issues ranging from function to ergonomics. Anecdotal experience based on one or two examples can give you a false read and, more often than not, seems to be the basis for unswerving opinions. A large inventory of well-used weapons provides more reliable data and is the source of most recommendations included herein.

It must be easy to operate. Disparate function is never a good thing during stressful circumstances and operational proficiency is commensurate with training. The more complicated something is, the more training will be necessary. If time and range access are issues, simple is better. Also, fewer parts offer less opportunity for breakage. But, even the simplest systems will quit working at some point, so…

Parts must be readily available. At one point well in the past, a cheaper series of Remington Model 870 pump shotgun clones were procured by my agency. Life was good for several years, until the manufacturer ceased production. I inherited the ensuing headaches associated with trying to source necessary replacement parts. Eventually, a large number of our shotguns were rendered inoperable. We finally switched to a reliable inventory of actual Model 870s, which are but a few of the 10 million-plus that have been in continuous production since 1950. We've never regretted this decision. Using a proven system, you can predict which parts are likely to fail and stock up on spares. In the case of our venerable pump guns, the list is fairly small.

Ammunition must be widely available. Just about any hardware store with an FFL will have 12-gauge shotgun shells on hand. The smaller 20-gauge is another fairly safe bet. If you need to scrounge, tip the odds in your favor. When you're buying, mainstream choices will just about always be cheaper. You'll also benefit from a much wider selection of loads to match your specific requirements. The same could not be said of 16 or 28-gauge shells, both of which are interesting choices, but have a limited following.

It must be easy to maintain. Function is crucial in a survival situation, so don't forget the maintenance factor. No firearm is truly weather-proof, and should be serviced after exposure to the corrosive effects of snow, rain, or salt water. Toolless disassembly really helps and promotes regular maintenance. In the case of our M 870s, it's reassuring to know we can breeze through a whole rack of guns in fairly short order, pulling off trigger groups, barrels, forends, and bolt assemblies for field servicing.

It should accommodate practical accessories. It's easy to overlook some simple necessities, one good example being a sling. Without one you're pretty much operating one-handed during any movement on foot. I was in the process of ordering a new shotgun and one of the first things I looked for was a set of QD sling-swivel mounting points. Interchangeable choke tubes are also high on my list (which we'll cover in depth). Depending on personal preference, a light rail, gun mounted shell carrier, or extended magazine may be useful. By sticking with the most popular guns, availability of such accessories is assured. Competitive product development further improves user convenience, with benefits like gunsmith-free installations. The downside is an increasing tendency toward over-accessorizing. One current fad involves turning just about every shoulder-fired weapon into an AR-15 clone. In the case of a shotgun, add-ons need not be excessive.

It must represent good value. The well-known rule of thumb is to buy the best equipment you can afford; however, when it comes to firearms, "equipment" means more than just a gun. Looking at the shotgun, some basics extras like a sling, case, choke tubes, ammo, and possibly a spare barrel will be needed. Adding up the essentials creates a dollar figure constituting the real bottom line. We call this our "system cost." Budgeting may be required to stay within our means, so an honest gun at a fair price helps keep a lid on costs.

There are plenty of good shotguns to choose from. A final choice can be difficult but, by keeping these principles in mind, we can narrow the field. A look at the various gun types may also help sort things out.

Chapter 2

SHOTGUNS

One of my instructors shoots a shotgun worth nearly as much as my late-model truck. It's a specialist's instrument designed for the game of skeet. He's an artist with this over/under, winning state and regional championships. In case you're wondering, it has no general utility whatsoever. I mention this to clarify that cost alone is not our criteria. We don't want junk, but we won't need to break the bank in order to obtain a reliable shotgun.

An expensive, high-end tournament gun. Fortunately, we can spend less!

ACTION TYPES

In firearm terms, an "action" houses the internal parts that make it work. This term also is used to classify general firearm types according to several prevalent designs. Let's take a look at some common types.

Single-barrel, break-actions. Such guns typically have hinged actions. Swinging a top lever or depressing a tab disengages the locking system. The front end of the barrel tips downward and the rear end elevates for extraction and loading shells. Many have exposed hammers that must be manually cocked for firing. Caution must be exercised when lowering the hammer if the shot isn't taken.

The overall design is very straightforward, and therefore reliable. You'll see some very expensive tournament guns on regulation trap ranges, but most are lower-priced, utilitarian tools. They're often marketed in smaller gauges for beginners. Many lean in the corners of barns or camps. Some are available with extra rifle-caliber barrels that can be easily switched. Others are offered in large gauges but, owing to their simplicity, most are light. A by-product is lots of recoil. Another downside: you only get one shot.

I think these guns have a place, but I wouldn't make one my primary choice. They're affordable and can be easily disassembled for storage, making them a stash-able option. Two or three-barrel combination packages afford a companion rifle option in centerfire and/or rimfire calibers. They're very simple to operate, which is a good thing in low-light environments or during stress ; however, those with external hammers will need to be un-cocked at some point. Novice shooters and external hammers aren't the greatest combination (although lots of folks may have a different opinion). You certainly shouldn't cock one until you're ready to fire, and you will need to be very careful when lowering that hammer. As for recoil, you can reduce it a bit through careful shell selection. Winchester sells a 12-gauge low recoil/low noise "Feather" load that behaves like a modest 20-gauge.

The smaller gauges generally offer less recoil, but gun weight is always a factor. I watched my young son get hammered by a fly weight .410. That happened only once, since I didn't want to see him develop a nasty flinch.

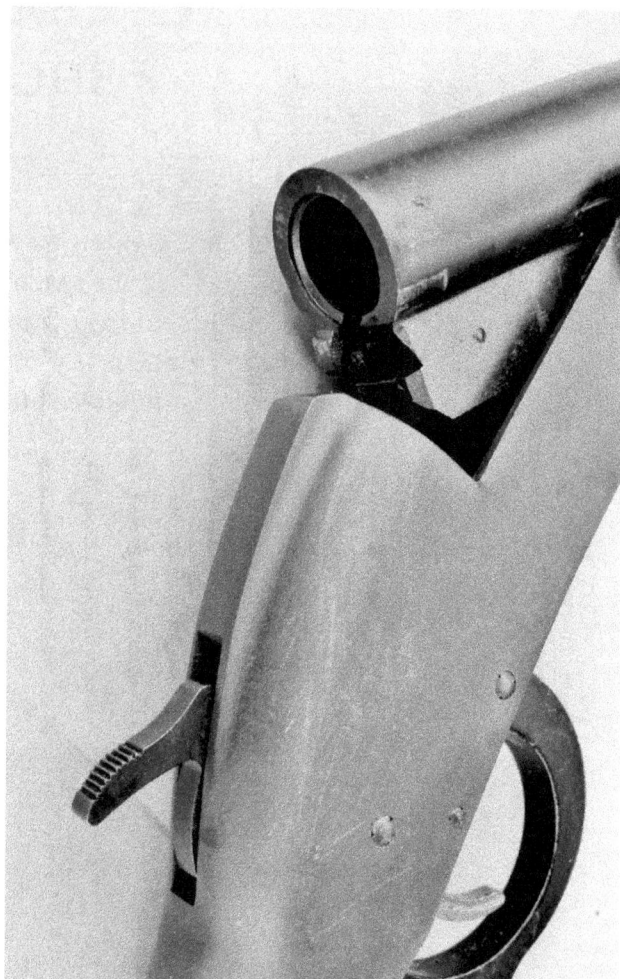

A single-shot break-barrel shotgun.

One neat thing about a break-barrel single shot is its ability to use small-gauge and metallic cartridge adapters. Essentially, these units are hollow, shell-shaped metal sleeves which can be inserted directly into the chamber.

Double barreled shotguns. Extending the break-action single-barrel design to a pair of parallel tubes creates this tried and true arrangement. Normally, the term "double barreled" refers to a side-by-side design (SxS), but the more popular over and under configuration (O/U) stacks barrels vertically.

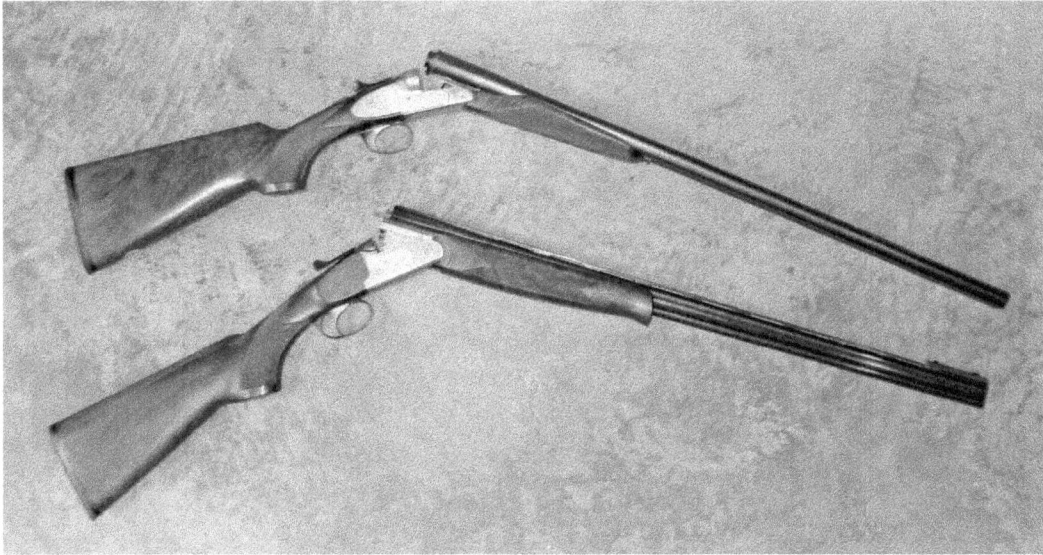

SxS (top) & O/U double-barreled shotguns, shown with actions opened.

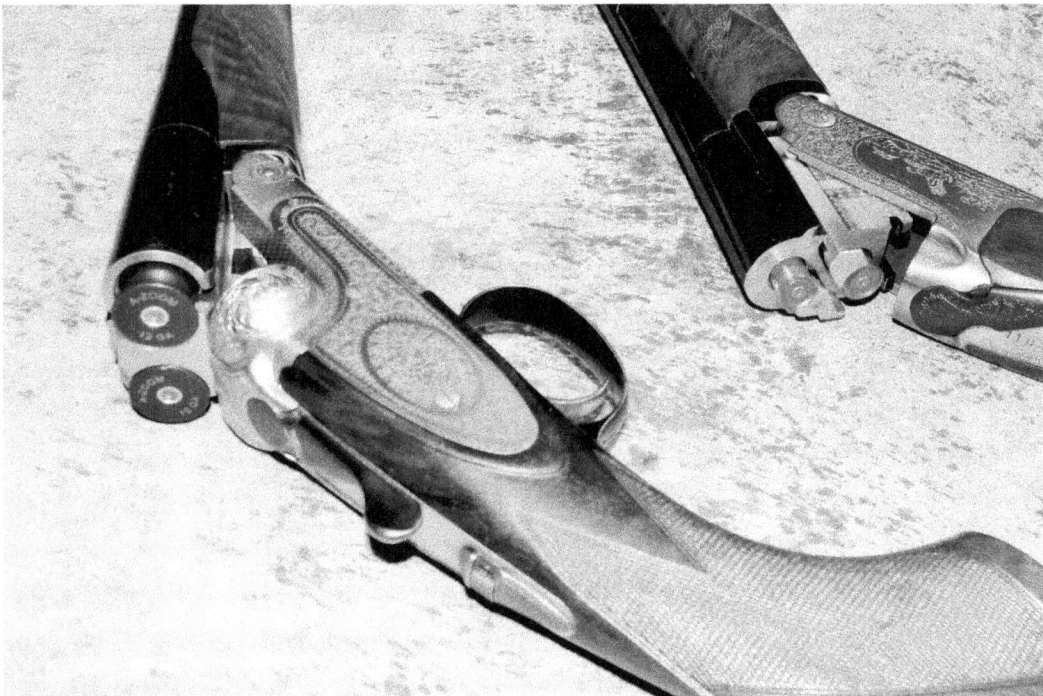

Close-up of SxS and O/U actions. Shells are inert "snap-caps."

Each barrel can be fired with its own independent lock work and trigger, or the firing sequence may be initiated with two separate pulls of a single trigger. In the latter case, the first pull normally fires the right barrel (side-by-side), or the lower one (over and under). The forward trigger on a double-trigger gun typically does the same thing. Nowadays, hammers are internal and automatically cock when the gun is opened. Shoving a safety forward will ready the gun for firing   .

A typical double-barreled shotgun equipped with double triggers.

We'll examine "chokes" later on. For now, applying slight degrees of constriction to a shotgun's muzzle can tighten patterns. Two barrels permit an opportunity for two fast choke selections. The first shot is generally from the more open choke, which will throw wider patterns. Double triggers (DT) provide the means for instant selection, and are often preferred by SxS traditionalists. A single, selectable trigger (SST) will have the means to change the firing sequence. Non-selective, single triggers (NSTs) don't provide this option, but are generally more reliable – an attribute worth considering with a less expensive gun.

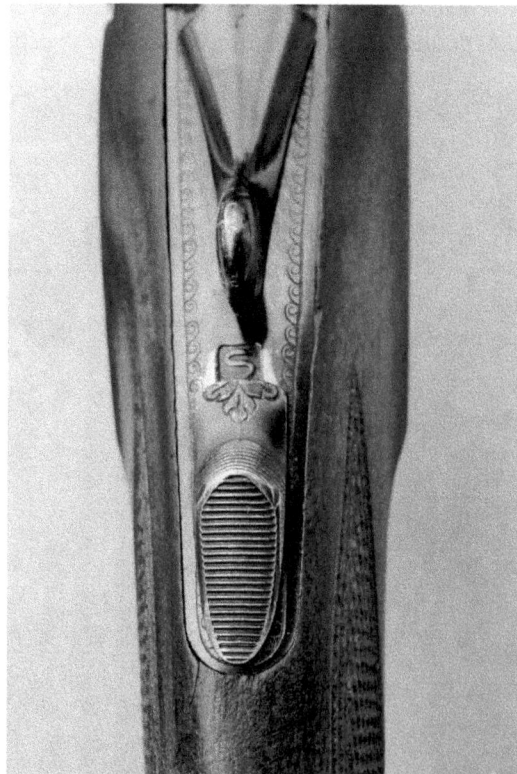

Opening lever & NST safety button. The right barrel will fire first.

Just about all double-guns have a tang-mounted (on the stock behind the action) safety button located behind the opening lever. The barrel selector is frequently related to the safety. It's a natural spot for your thumb to find, and works with either hand. Often, field guns will be equipped with "automatic safeties" that mechanically reset when the gun is opened. Some folks like this feature and others don't. I prefer manual safeties for quick, sustained firing.

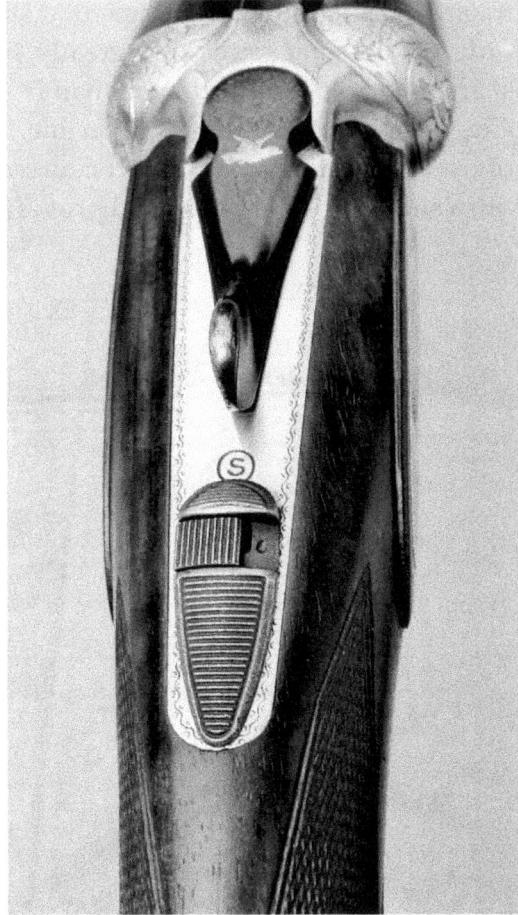

This gun is equipped with a SST, shown on "safe", set to fire its right barrel first.

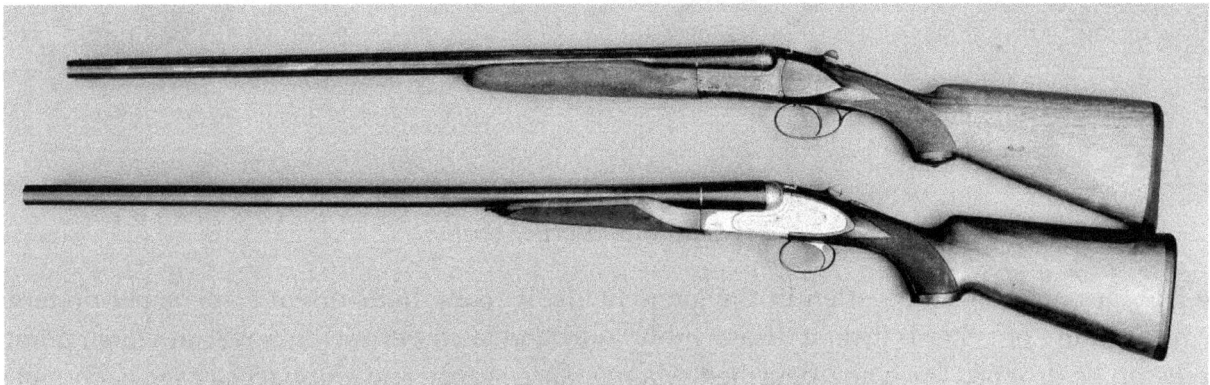

A pair of single-trigger Italian double-guns in 12 & 20-gauge chambering.

A good side-by-side can set you back several thousand dollars, and will likely be of European lineage. I've played with much cheaper Turkish imports to varying degrees of success. The South American guns I've handled had the grace of a truck axle, but would serve for home defense. In fact, some are marketed accordingly, with light-mounts and short barrels. Their virtue is simplicity, and looking at a pair of muzzles should get anyone's attention.

Over/unders are far more common. The single sighting plane of a stack-barrel is frequently touted as an advantage over a side-by-side. Quality and prices vary greatly, but overall, you can buy a good O/U for much less money. Most are made offshore and many hail from Europe or Japan. Two dominant brands are the Browning Citori series, and Beretta's M-686 line. The Ruger Red Labels are built in the U.S.A. You'll see plenty of over and unders on claybird courses because they handle well, afford two ready choke options with a single sighting plane, and provide dependable service - qualities also recognized by hunters.

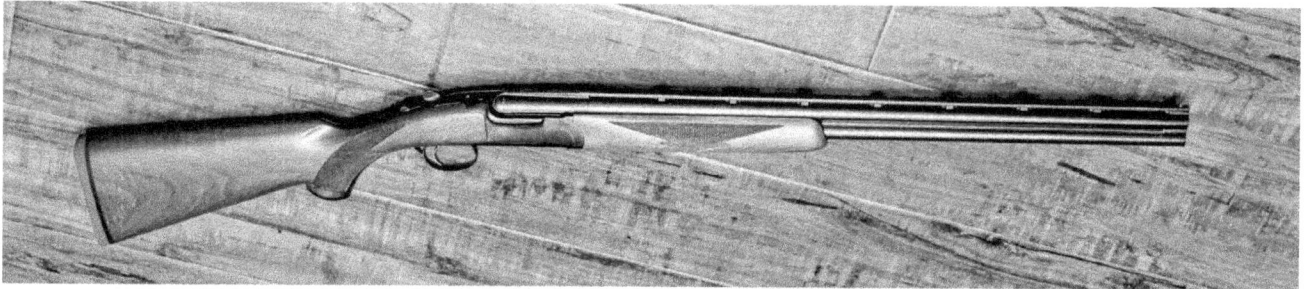

A nice, original Ruger Red Label O/U, chambered in 20-gauge.

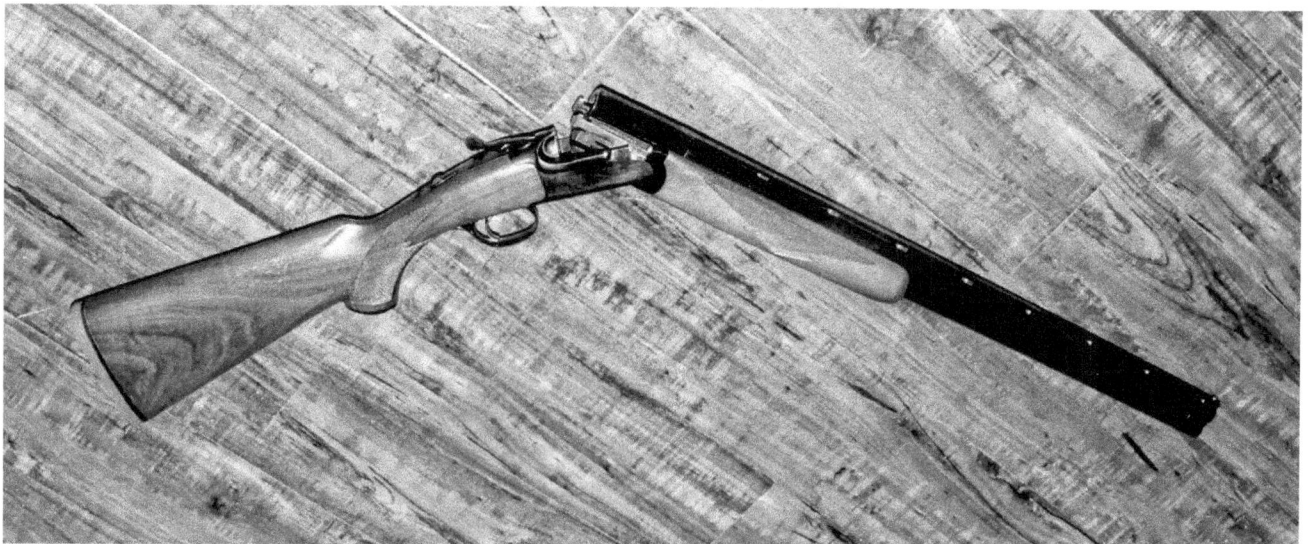

Red Label with its action open.

Side-by-sides are seen more often in the game fields. It really boils down to personal preference. One nice feature of either is instant disassembly. A forend latch permits removal, at which point the barrels can be dropped open and detached. Maintenance is easy and compact storage is possible for convenient transportation.

For either system, the disadvantages are higher cost, the two-shot limitation, and difficulty firing slugs with accuracy. I really like my two-barrel guns, but if survival purposes are your objective, as much as it pains me to say this, they're not my pick.

However, America is full of old double-guns, many having been handed down through genera-tions. Suitability, quality and condition vary greatly. I'd take an old-timer to a gunsmith prior to touching one off. If it works, you have a shotgun. It'll probably have plenty of stock drop, two trig-gers, and lots of recoil. Pulling both simultaneously will GUARANTEE lots of thump. Most of the old guns are tightly choked and throw dense patterns with modern shells. An ill-fitting gun won't shoot where you're looking, and will beat you to death. As a teenager, I once knocked myself into next Tuesday by inadvertently pulling both triggers at once with an old 12-gauge double-gun. I only did that once.

On the other hand, the previously-mentioned inexpensive South American guns are robust and sim-ple, if not graceful. A 12 or 20-gauge has potential as a standby defense gun, and is as easy to operate as a single shot. The tang-mounted safety is ambidextrous. The Picatinny rails on the short-barreled, home-defense versions permit attachment of lights. After things calm down, they're easy to unload. With the action open you can verify a cleared gun, even in the dark, by poking each chamber with a finger.

I really enjoy using double-guns. They're simple and relatively safe. As an upland bird hunter, obsta-cles are always an issue. Old wire fences, stone walls, and ditches are safely navigated by opening the gun and plucking out both shells. It only takes a couple seconds and promotes user longevity. Plus, there's just something special about a graceful, well-built double, and esthetics do count for some-thing. Trench warfare isn't one, explaining why some of our doughboys went to the front, armed with Winchester Model 97, pump action shotguns.

Slide-action (pump-action) shotguns. While the break-action doubles have offshore roots, the pump is a grassroots design, firmly entrenched in American culture. Even those with expensive double-guns are likely to have a pump on hand for rough and tumble gunning. It's the go-to gun for just about every American sportsman. Many squad cars have a shorter "riot gun" version on board, and the military uses the same system. Why? Because they work!

The forend is connected to a strong breech bolt. It reciprocates with muscle power on a magazine tube hanging under the barrel. As the forend and bolt travel rearwards, an internal hammer is cocked, and a shell pops out of the magazine. On the forward stroke, it's elevated by a carrier, and shoved into the rear end of the barrel by the bolt face. With practice, this process can occur *really* fast! Most tubular shotgun magazines hold 3-5 shots, and extensions are available to hold a few more. The increased capacity and speed of operation translate to increased firepower.

Manual operation ensures function with a large variety of shells. Unlike semiautomatic designs, no minimum shell power level is necessary. I'm more of a semi-auto fan, but our agency's troops are equipped with pumps for a reason. Besides conventional buckshot and slugs, our trusty slide-actions will work with specialty shells like bean bags, which generate very low pressure.

Because the receiver houses a reciprocating bolt, overall gun length will be several inches greater, compared to break-action gun. A 26" barrel pump will have an overall length similar to a 29" double-gun, but will provide at least twice the capacity.

The manual-of-arms is a bit more complicated, but still fairly straightforward. We tell our shooters to run the gun with authority, which promotes positive feeding and ejection. This is seldom a problem for an able-bodied man, but might pose problems for kids or women with shorter a reach. Careful shopping should solve this problem.

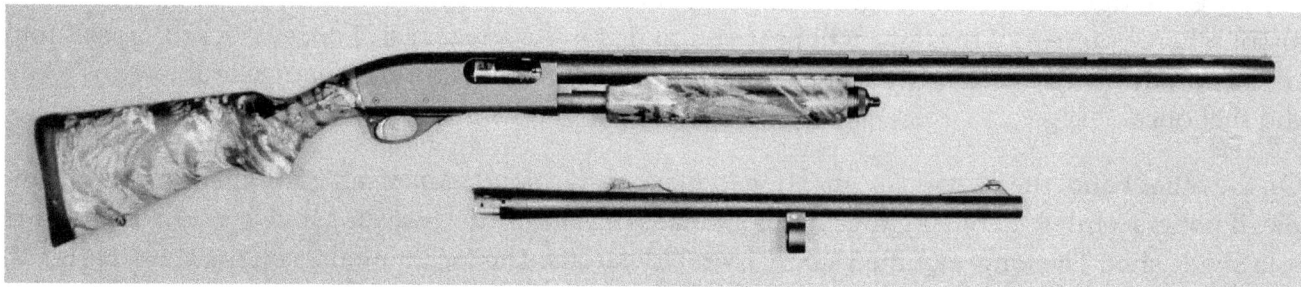

A switch-barrel sporting pump-gun based on the venerable M-870.

Many pumps can be easily broken down for compact storage, or to permit a quick barrel change. We have a 28", ventilated rib Remington Model 870 on hand at the house. It's more or less a do-all gun that sometimes equips a guest hunter. Interchangeable choke tubes cover just about all wing shooting needs and it's taken a fair share of turkeys. By simply unscrewing the magazine cap, we can switch out the bird barrel for a handy, 21" rifle-sighted slug barrel and hunt big game. Such versatility is hard to beat.

Semi-automatic shotguns. Function is somewhat similar to a slide-action, but no pumping is required. Recoil energy or combustive gas pressures provide the impetus to drive the breech-bolt rearward when a shot is fired. Once compressed, a strong spring shoves the bolt forward again, pushing a shell into battery. Ejection, feeding and chambering all occur during this carefully timed process. Unlike a pump, shells must provide power levels compatible with the gun's design. Engineers have been tinkering with this principle for more than a century and as a result, today's systems are surprisingly reliable.

The prevalent design is the so-called "gas gun", which diverts propellant pressure through small gas ports. The gas pressure has enough force to drive a piston rearward and power the bolt. One benefit of gas-operated shotguns is a noticeable reduction in perceived recoil. Between gas pressure taking a 180-degree turn, movement of the piston, operating rods and bolt, function is staged. An incremental effect occurs during recoil, which really helps soften heavy loads. A proliferation of magnum shells posed challenges for manufacturers, who have overcome most early problems with clever pressure metering designs. I continue to be amazed by the variety of loads a good gas gun will digest. I'll also hear an occasional complaint about reliability. Five will get you ten it's related to maintenance; or lack thereof. By nature of its design, the system injects combustive residue into the action. Bottom line: folks just don't clean them. I've seen bolts slide in slow motion as a shell was manually loaded.

Amazingly, most of the time these neglected guns still ran. Regular cleaning and occasional spring replacement, combined with a proper diet of shells, will keep yours running all of the time.

Gas Ports

High-pressure combustive gas is channeled through two barrel ports.
It can be harnessed to power a semiautomatic action.

Recoil-operated shotguns are a different breed. Instead of gas pressure, they harness momentum, which normally involves the barrel running rearward to cycle the bolt. It's an early design, perfected by the brilliant dean of firearms design, John Browning. You won't get the fouling seen with a gas gun, but the absence of a metering system can narrow shell choices. Function is sometimes adjustable using reversible "friction rings", which compensate for various recoil forces. Speaking of which, a certain product is just that – recoil. These guns tend to be lighter, thanks to the lack of pistons and related parts. They handle well, but without a staged function process, and with that barrel mass running rearward, you'll feel a noticeable recoil increase compared to gas guns.

The latest rage is the "inertial system", also relying on recoil-generated momentum; but in this case, a rotating bolt head unlocks from the barrel when a separate, larger bolt component begins rearward motion. It's an ingeniously simple design that runs clean, and functions reliably in nasty weather. The spectrum of acceptable loads is sometimes narrower, but waterfowlers feed them magnums by the case, having little interest in powder puff shells. Cleaning an inertia driven shotgun is a revelation for anyone schooled on gas systems. There's not much crud, and fewer parts to mess with. You may feel more recoil, but recent stock designs with built-in shock absorbers have helped.

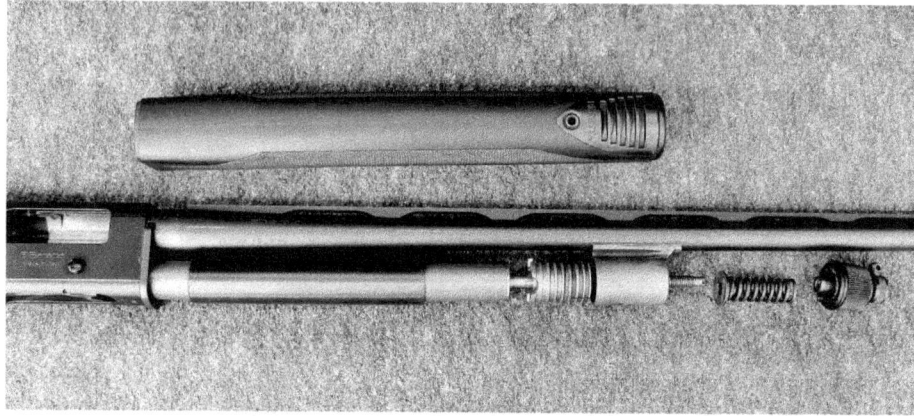

Under the hood of a gas gun: note the piston that drives the action rearward.

I like the gas guns. Believe it or not, the ability to withstand recoil is not some test of manhood. Heck, I work on a range and lots of shooting goes with that turf. Trust me, you'll shoot a whole lot better if you're not picking your head off a stock, closing your eyes, flinching, or just as likely, doing all of the above. Combined with the right loads, gas-operated shotguns do noticeably soften recoil. Believe it or not, a properly fitted 20-gauge autoloader is a great way to start a new shooter. You can begin by loading single shells, gaining the safety of a single shot without the recoil you'd get from a lightweight break-barrel. You won't have an external hammer to deal with either.

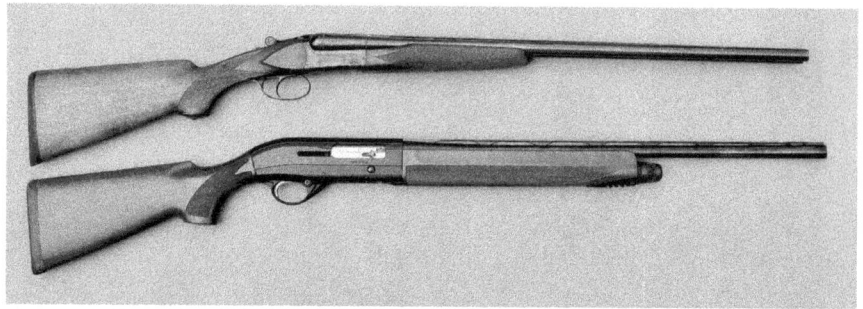

A pair of 20-gauge shotguns: the semi-auto gas gun has a 24-inch barrel and will have less perceived recoil than its 26-inch side-by-side companion. Note their similar lengths.

We do a fair amount of crow control, which involves high volume 12-gauge shooting. Our gas guns really save the day as we pound through several boxes of high-velocity shells. I watched my son get pretty numb shooting the same loads in his short over and under. Plenty of people shoot sporting clay courses with gas-operated autoloaders for exactly this

Crow control with a Remington M-1187 gas-operated semi-auto shotgun.

reason. Admittedly, cleaning is a chore and regular maintenance is advisable. I'll gladly pay that price. Meanwhile, the continued evolution of inertial designs is causing me to rethink everything. I shot the latest Beretta 1301 Competition gas gun against an inertial Benelli M-2 and both handled well. During subsequent cleaning the Benelli looked largely unfired!

Oddball designs. There are different designs that are sliding into obsolescence while others have just appeared. In the former category would go lever actions like the old Winchester repeating M-87, and single shot break-barrel Ithaca. A run of .410 Marlin and Winchester lever-guns appeared recently, but was short-lived. Rossi filled that vacuum with their Rio Grand, a Marlin M-336 knock-off. Kel-Tek's unusual Bullpup Model KSG twin magazine pump is among the latest designs. We now have a 12 Ga. AR-15 clone, and even AK-style shotguns.

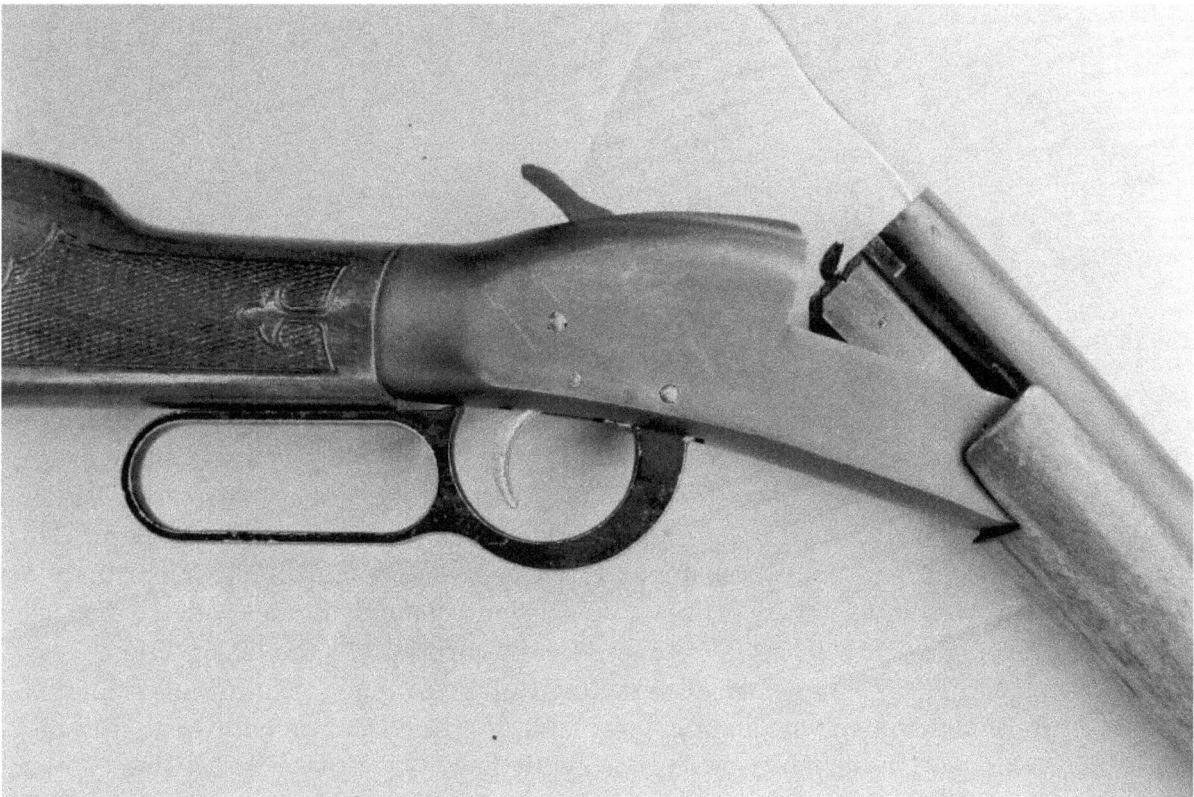

An odd-ball design: Ithaca's old lever-action single shot Model 66 shotgun.

Chiappa even offers a three-barrel variation of a double barreled shotgun, which locates the third barrel above and between the bottom pair. It's not really a new idea, but does represent a modern spin, and is offered in sporting or defensive-length versions.

In the 1960s, inexpensive repeating bolt actions were fairly common, and filled a bargain basement niche. The surviving offspring are specialty slug guns with rifled barrels for use in shotgun-only areas. They're really rifles, and offer surprising accuracy beyond 150 yards. Lots of older bolt guns are still around, but one wouldn't be my first choice, or even second. On the other hand, if you already have one, it'll make a good spare shotgun.

A typical low-cost bolt-action shotgun from the 1960s.

A new phenomenon is the revolving shotgun, configured either as a handgun or shoulder-fired affair. Taurus came out with the Judge, which is a long-framed revolver. It has a rifled barrel and swing out cylinder, accepting .45 Long Colt cartridges, or .410 shotgun shells interchangeably. The rifling is required to avoid BATF sawed off barrel shotgun classification. The combination of anemic .410 loads and wider patterns (dispersed by rifling), limit use to close range. Accordingly, the primary market is self-defense, and some interesting new buckshot, slug, and combinations of both have appeared to meet this need. During recent but casual testing, we were surprised by the major differences in patterns from one type to the next; however, even with the top performers, effectiveness is limited. Next came Rossi's Circuit Judge revolving shotgun, available as a true shotgun with an 18-inch barrel, in .410 or 28-gauge. Smith and Wesson came on board with their .410 "Governor" revolver. It even fires .45 ACP cartridges. Although these guns may meet specialized needs, for general utility it's hard to beat a conventional gun.

That's where this pathway will lead us, but we should stop along the way to sort out the sometimes confusing business of gauges and shotgun shells.

CHAPTER 3

AMMUNITION

If ever something was steeped in tradition, it's got to be the nomenclature associated with shotgun shells. Since a shotgun is a smooth bore designed to fire multiple projectiles, there is no critical bullet-to-barrel fit. Still, *some* degree of standardization is necessary. Many related shotgun principles date to the industrial revolution, or even earlier.

SHOTSHELL COMPONENTS

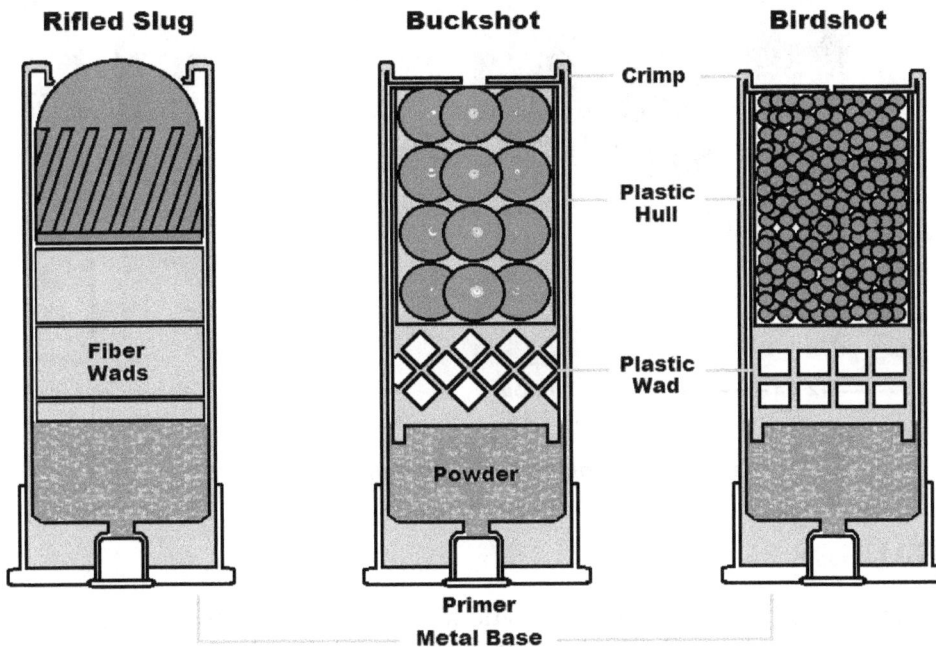

The anatomy of three common shell types.

GAUGES

A 12 Ga. is bigger than a 20 Ga., which might seem backward. It's an old British system that "gauges" a shotgun barrel. The gauge number is based on the number of lead balls matching its inside diameter that will equal a pound. If we nudged a lead ball through a 16-gauge barrel, we'd discover each one weighed an ounce. If we set our scale to balance out at one pound, we'd count 16 of these balls. If larger barrel was gauged with this system it would take fewer, heavier balls to hit the 1-pound mark. Examining a 12 Ga., each ball would weigh 1 1/8 ounces and it would only take 12 of them to equal a pound. On the other hand, the smaller 20-gauge would require more lead balls to reach the pound mark. Because they're smaller, each of the 20 lead spheres would only weigh 7/8 ounce. A 28-gauge is even smaller and a properly proportioned twenty-eight is a joy to carry. An anomaly is the little .410 bore, which isn't a gauge at all. At the other end of the spectrum lies the mighty 10-gauge that was very popular a century ago, gradually losing ground to the now reigning twelve.

The ten has seen resurgence with waterfowlers because it can throw more pellets. Federal law prohibits lead shot for ducks and geese. Many hunters fire lighter steel shot, which hits with less punch. More pellets can help offset this drawback so the larger gauges are the standard choice when decoys and marsh grass is involved. The tiny .410 is often chosen to get kids shooting. I think that's a mistake. Its small shot charge requires precise shooting, so a carefully chose, youth 20-gauge would be a better pick.

Common American shotgun shells in magnum & standard lengths.

COMMON GAUGES

Note the "standard" and "magnum" lengths. The longer magnums hold more shot but Newtonian laws dictate that they'll also produce more recoil. The term "magnum" is often used to describe loads with souped-up performance and cartridges are often lengthened versions of less potent, original versions.

Gauge Standard Magnum Availability Use
10-gauge 2 7/8" 3.5" Limited *Primarily a waterfowl choice for use with heavy, non-toxic shot loads. Shells are expensive. Guns are heavy and so is recoil! The shorter shell is now obsolete.*

12-gauge 2 ¾" 3"/3 ½" Most popular *Covers everything, with a huge variety of shells from light target loads, through 3" Magnums. The newer 3 ½" Super magnums designed to approach 10 Ga. performance.*

16-gauge 2 ¾' N/A Semi-obsolete *A well-designed performer, now supplanted by 20 Ga. 3" Magnum offerings. Shells are expensive and harder to find, but are still used by a small group of loyal bird hunters.*

20-gauge 2 ¾" 3" 2nd most popular *Comes in right behind the 12-gauge, offering less recoil with useful loads in smaller guns. Popular with youth and ladies. Magnum shells increase punch, but also recoil.*

28-gauge 2 ¾" N/A Uncommon *Enjoys a cult-like following by upland bird hunters. Also used on skeet fields. Approaches 20 Ga. in perfor+mance but shells are less common, and costly.*

.410-bore 2 ½" 3" Limited *Small shot charges severely limit performance. Sometimes used for beginners. Used on skeet fields. You'll gain small guns, but hefty pricing for ammunition.*

There are others, including some pretty obscure offerings such as 24 or 32-gauge. For our purposes, availability is a primary concern, so only the most common types will be examined.

10-gauge. The largest common offering, use is mostly confined to waterfowl with heavy, non-toxic shot loads. Predator hunters sometimes buckshot to stretch their reach on coyotes, and the occasional turkey hunter will use large doses of conventional-sized pellets for the same reason. Shells are big and expensive, guns are heavy, and so is recoil! Surprisingly though, the mass of the big guns may result in less recoil than you'd receive firing 3 ½-inch 12-gauge super magnums.

> ➤ 2 ⅞" *Standard:* This load is now obsolete, meaning that the new "standard" is actually a magnum.

> ➤ 3 ½" *Magnum:* The huge payload capacity will handle 1 ¾ ounces of steel shot, or 2 (or more) ounces of lead. You can throw twice as many 00 buck pellets (18) as you can from a 2 ¾" 12-gauge.

These numbers are impressive, but I'd skip the big 10 in favor of a do-all 12-gauge unless you're built like The Incredible Hulk, and already have a wagon load of shells.

12-gauge. The most popular choice, available in a huge selection of loads from low recoil target shells to screaming super magnums that approach the performance of the big ten. Most folks of average build can handle this gauge, including many women and teenagers. The key is proper load and gun selection.

> ➤ 2 ¾" *Standard:* You can start with ultra-light loadings like Winchester's "Feather Low recoil/Low Noise" subsonic shells that are milder than a ⅞-ounce 20-gauge (one caution: they may not cycle a semi-auto). Moving up a notch, most 12-gauge target loads throw 1 to 1 ⅛ ounces, producing effective results on claybirds without excessive recoil. Stiffer "high brass"

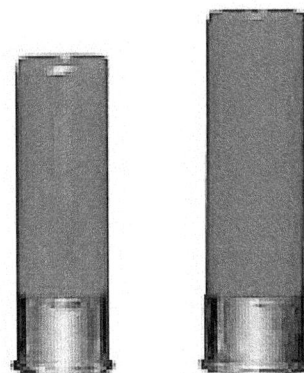

A 10-gauge shell (right) beside a 12-gauge.

(or high-base) loads toss heavier payloads of 1 ¼ to 1 ⁵/₈ ounces for use on game. The heavier field loads usually have taller brass heads, which don't always equate to more power.

➢ **3" Magnum:** The longer hull holds more shot. It also kicks more. Many employ it where a lesser load would suffice, but there's no denying a magnum's effectiveness on tougher targets. The extra room permits more pellets, which can offset the reduced effectiveness of lighter steel shot. Lead pellet loads can run up to 2 ounces.

➢ **3 ½" Super magnum:** A recent development, this wild looking shell was bred to deliver copious quantities of steel or other non-toxic pellets federally mandated for waterfowl. I still remember the first 3 ½" I touched off, which was from a fairly light pump gun. It really rattled my cage! A softer autoloader helps and manufacturers have made headway improving function with 3" and 2 ¾" shells. I shoot with a couple knowledgeable gun people who use 3 ½" self-loaders, but they mostly burn lighter shells. Surprisingly, their guns even digest 2 ¾" trap loads. Their argument is that it's nice to have the extra punch available if ever needed. A full house dose of lead shot weighs 2 ¼ ounces – nearly the same as a 10-gauge!

Common 12-gauge shells in 3 ½; 3 & 2 ¾-inch lengths.

If limited to just one gun, a 12 Ga. should be your first pick. In fact, I'd have at least one regardless of how many guns I owned. In a pump or auto, I'd be sure to select a 3" Magnum. There are so many great 2 ¾" loads available that you'll seldom need this punch, but it's nice to have the three inch option for waterfowl, turkeys, or larger critters.

I'd skip 3″ Magnums for self-defense in order to improve recoil control and reduce muzzle flash. We run several law enforcement shotgun programs weekly, affording plenty of opportunity to watch he-men shoot. We also see plenty of misses resulting from flinching. This is with low recoil, tactical 00 Buckshot inside 25 yards. Things were worse back in our full-power buckshot and slug days.

The 3 ½″ loads kick like the Dickens. Unless you're a highly experienced shooter with special long-range requirements, I'd stick with a 3″ gun. During a goose hunt on Maryland's Eastern Shore I asked our veteran guide what he preferred. His go-to choice is a good 3″ Magnum #2, fired from a gas-operated autoloader, which softens recoil. I brought one for the same reason and didn't feel handicapped at all.

A limit of Eastern Shore Maryland geese shot with 12 Ga. autoloaders.

16-gauge. It's a whole lot more popular in Europe nowadays than it is in the United States. Too bad, because the sixteen is a well-balanced design with a shot column proportionate to its bore. Magnum 20-gauge loads really did it in, approximating the sixteen's performance in a smaller gun. The standard 16 Ga. load is 2 ¾″. Guns built on true 16-gauge frames are a bit trimmer than their larger cousins, and you will still maintain useful hunting punch. Payloads run up to 1 ¼ ounces; however, for our purposes I'd skip it unless you already have a gun you can't live without. Even then, shells will be harder to come by and cost may be double what you'd pay for garden variety 12 Ga. loads.

A 16-gauge shell (center) framed by 3" Magnum & 2 ¾" 20-gauge loads.

20-gauge. This is the second most popular gauge, falling in right behind the twelve. Shells and guns are considerably lighter if the latter is built on a dedicated frame. With reasonable weight and dimensions, the twenty is a popular choice for ladies and young shooters, offering less recoil as a bonus. That said, I've fired some lightweight 20-gauge guns that produced stiff recoil. Fire lighter target loads from a gas-operated autoloader and you'll have a very pleasant experience. I really enjoy the twenty and use it extensively for upland bird hunting. A well-proportioned gun is a joy to carry and I've seldom found this combination lacking. Where more oomph is called for we can choose some surprisingly effective loads.

➢ **2 ¾" Standard:** Quite literally, tons of claybirds are disintegrated on a daily basis with $^7/_8$-ounce target loads. The same shells will do the job on many upland birds. Hotter loads toss up to 1 $^1/_8$ ounces of pellets, and can tackle tougher quarry like pheasants.

➢ **3" Magnum:** Here's the load that darned near killed the great sixteen. You get similar performance in a smaller package, although some will argue the twenty's narrower shot column throws more fliers for any given weight. Regardless, the magnum twenty works. It holds enough pellets (1 ¼ ounces) to be useful within reasonable distances, and the latest sabot slug loads are downright nasty.

Like the twelve, I'll pick a 3" chamber whenever possible just to gain added versatility. I seldom shoot the magnum shells, mainly because a 12-gauge is never far away, but it's nice to have that option. Using side-by-sides afield, I'll often load the first barrel with a $^7/_8$-ounce target shell. The second chamber may contain a stiffer one ounce game load. For jump-shooting wood ducks I might switch to 3" non-toxic magnums.

The twenty-gauge is a great loading that certainly has a well-deserved reputation. I'll always have at least one on hand. For those unable to handle a larger bore it's the logical choice; however, for any one-gun-only shooter capable of handling it, a twelve is the better pick.

28-gauge. This smaller shell has a cult-like following among dedicated upland hunters. A shotgun built on a properly scaled frame is a joy to carry. Performance approaches the larger twenty, but shells are expensive and often hard to find. The standard load is ¾ ounce, but a one-ounce load is available. Brenneke loads a ⅝-ounce rifled slug as well. Even though the neat little twenty-eight is well respected, ammo issues preclude it from serious consideration for most use; however, a recent curve-ball is Rossi's revolving shotgun, the Circuit Judge. Picture a large double-action revolver with a shoulder stock and 18-inch smoothbore barrel. It could conceivably serve as a do-all, close range meat or self-defense gun, assuming a large supply of shells were stockpiled. You'd probably pay more for those than the gun.

An efficient 28 Ga. shell beside 3" Magnum & 2 ½" .410-bore loads.

.410-bore. Notice it's listed as a bore, instead of a gauge. The "bore" means caliber, and .41 is small for a shotgun. Still, I'll admit to a soft spot for the tiny .410. For me it's mostly a toy. My svelte Rizzini over and under spends most of its time in the nether regions of the safe, but once in a while, it emerges for some entertainment involving either clay birds or woodcock. My shooting partner, Mike, had a German Shorthair that was the ultimate pheasant pointer. We quickly learned there'd be little to eat, using more traditional ring-neck loads. So we broke out our .410s and had a blast, shooting three inch, #7 ½ shells. The birds were booted out of thick brush at contact distance and engaged inside of 25 yards. Mostly, it was just a stunt. Some people consider a .410 a beginner's gun, but its small payload requires precise delivery. We consider it just the opposite; an expert's choice. Two common shells are offered:

➤ **2 ½" *Standard*:** This little shell contains only ½-ounce of shot. Unless the pellets are small, not very many will fit. Shot sizes run from #4 through #9. Most 2 ½" loads are fired at clay pigeons during regulation skeet. A $1/_5$-ounce slug is also available, but it really isn't a good big game choice at all. Of late, some very unusual self-defense loads have appeared for use in .410 revolvers (we'll examine those in the separate handgun manual). The .410 shells, regardless of type, are really quite expensive. One solution is to load your own. I scrounge Winchester AA hulls at every opportunity and reload them with a basic MEC 500 Jr press. I don't shoot a lot of .410 and the small charge of #9 pellets saves on shot cost. Recoil is just about non-existent, and the entertainment factor makes the effort worthwhile. I owned a 10" .45 Long Colt/ .410 Contender barrel for a while. Hitting clay birds with a pistol was a challenge, although not impossible. After tagging out during archery deer season, I'd sometimes stake out apple trees for close range grouse hunting. Shots were mostly at stationary or walking birds, but things were still pretty tricky. In other words, the little shell most definitely has its limits.

➤ **3" *Magnum*:** The extra ½" increases shot capacity, resulting in an $11/_{16}$-ounce payload. The narrow shot-column isn't the best for optimum patterns, but field performance is nonetheless improved. Most 3" loads will contain shot sizes from #4 through # 7 ½ pellets. On the afore-mentioned pheasants, I ran #7 ½ in the bottom barrel and a #6 in the top. It worked within reasonable distances. Still, fairly precise shooting was necessary. The recipe for good hits involves well-grounded wing-shooting principles, whereby the gun is pointed, and NOT aimed. If you aim, you'll probably miss. Precise pattern-placement without this knowledge can encourage aiming. It's a vicious circle, which is why the .410 shouldn't be a beginner's choice.

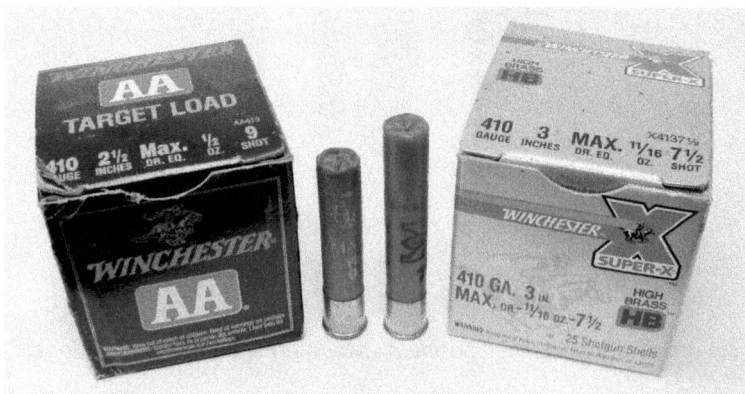

The shorter 2 ½" loads are often reserved for clay target games like skeet. The longer 3" shells are primarily used afield.

For the most part the smaller 28-gauge and .410-bore are really specialty choices. Shells expensive, sometimes difficult to find, and limited by lighter shot payloads. The practical cut-off for general shotgun chores stops with a 2 ¾" 20-gauge. Between the 20 and the 12, just about all bases can be covered. The one most likely to meet ALL requirements is a 12-gauge.

COMMON SHOTGUN PROJECTILES

The previously described lead ball "gauge" weights pretty much translate to the standard shot pay-loads for respective gauges. The term "shot" refers to spherical pellets which are expelled simultaneously upon discharge. The actual quantity depends upon several factors including pellet size and payload weight. A 20-gauge, 2 ¾" target load typically contains ⁷/₈ ounce of shot. The standard 16 Ga. load contains an ounce. A 12 Ga. 2 ¾" target shell normally contains 1 ¹/₈ ounces. In any given pellet size, the quantity will increase commensurate with the shot charge weight. The above numbers blur, due to "high brass" (or high-base) loads that hold additional shot. One ounce, 2 ¾" 20-gauge loads are readily available, as are 1 ¼ ounce 12 Ga. "field loads." More pellets can improve pattern density, increasing hit probability at longer ranges. But, everything comes with a price, which in this case is recoil.

Shot size. The American system assigns numbers to quantify pellet sizes. Like gauges, a lower number indicates a bigger pellet. A handy formula to determine actual pellet diameter involves the number 17. Coincidentally, that's the approximate size of a BB in hundreds of an inch (0.177), or caliber. If we subtract the number denoting a pellet's size from 17, we can determine its diameter in hundredths of an inch or caliber. For example, a #2 pellet is just a bit smaller than a 17-caliber BB, measuring approximately 0.15". It's among the largest birdshot sizes and is typically used on geese. A #9 pellet, measuring around 0.08", is the smallest common size and is used for short-range claybird games like skeet. Pheasant hunters sometimes prefer #6 shot, and plenty of them use #4. The more common sizes are charted below, but there are others like #8 ½ or #7 pellets.

Common American Lead Birdshot								
Pellet size:	**#9**	**#8**	**#7**	**#6**	**#5**	**#4**	**#2**	**BB**
Diameter:	.08	.09	.095	.11	.12	.13	.15	.177
Per ounce:	580	400	350	225	170	135	90	50

TARGET LOAD

AVERAGE PELLET COUNT							
OUNCE OF SHOT	SHOT SIZES				INTERNATIONAL		
	7½	8	8½	9	GRAMS OF SHOT	SHOT SIZES	
						7½	8½
1	350	410	497	585	24	296	420
1⅛	394	461	559	658			

AVERAGE PELLET ENERGY (FT-LB)							
DISTANCE (YARDS)	SHOT SIZES				INTERNATIONAL		
	7½	8	8½	9	DISTANCE (YARDS)	SHOT SIZES	
						7½	8½
20	2.1	1.8	1.4	1.1	20	2.4	1.6
30	1.5	1.3	1.0	0.8	30	1.7	1.1
40	1.1	0.9	0.7	0.5	40	1.3	0.8

DATA BASED ON MUZZLE VELOCITY 1200 FPS   DATA BASED ON MUZ. VEL. 1325 FPS

This box of target shells is a handy information source.

Lead shot. The traditional material for pellet composition has been lead. Pellets were formed by pouring molten lead through elevated screens that had sizing apertures. As the lead droplets fell from the top of a "shot tower", they became spherical and dropped into water, rolling down inclined planes to sort and grade them.

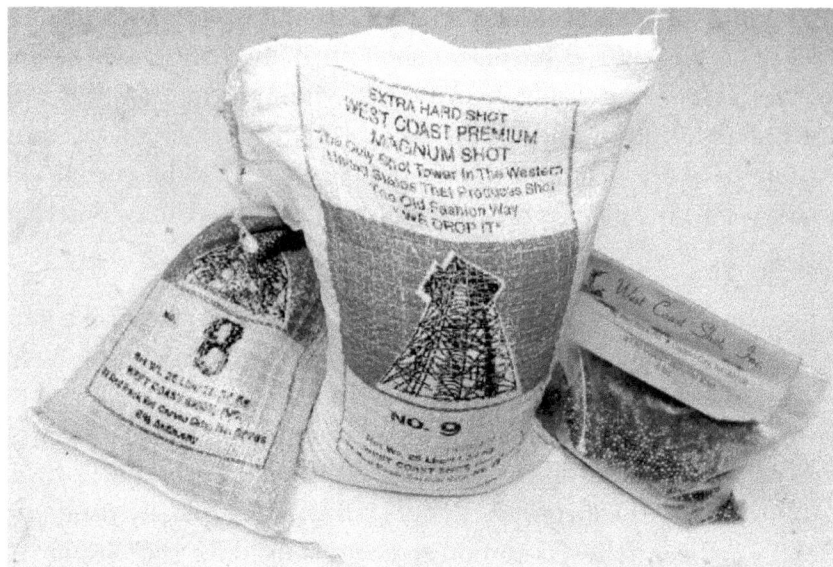

Lead and copper-plated shot. Note the shot tower logo on the bag of #9s.

Upon discharge, lead pellets are subject to violent forces, the first being inertial setback. The top part of the shot charge momentarily remains at rest while the bottom layers accelerate. Inevitably, some pellets will be deformed, which can cause irregular patterns. Further pellet damage can occur to those contacting the barrel walls. Next, the pellets encounter the barrel's choked portion at the muzzle. All of these collisions conspire to degrade pattern performance and great effort has gone into countering these effects. A simple fix is to harden the pellets by adding antimony. This alloy is expensive, explaining why good shells cost more. Coating the pellets with copper or nickel plating can further help with a corresponding increase in price. Cheaper shells use softer shot, which throws larger and more irregular patterns.

Old fiber and new plastic wads.

A shot charge needs some sort of seal behind it to capture propellant forces. During the 1960s, plastic wads gained traction, quickly replacing older fiber wads used to contain pressure. By adding thin sleeves to the wads, a protective cup could help isolate shot pellets from barrel and choke contact. Performance improved and gun makers began adjusting their choke constrictions to maintain established pattern standards (see the choke section). Next, granulated plastic buffering material was distributed among the shot charge to cushion individual pellets during discharge. When we buy so-called "premium shells", these improvements account for much of their cost; however, we may not always need them! For sandlot clay bird practice or close range upland birds, cheaper shells will work. Patterns may also be bigger, which can increase hits on difficult crossing targets.

Non-toxic shot. Steel shot is sized to the same grading system, but larger pellets are more commonly preferred. Most use is geared toward waterfowl, due to a federal prohibition of lead. In a given size pellet, steel will have less terminal energy than lead simply because steel is lighter. The remedy involves switching to larger sized steel pellets. Back when lead was legal, I liked #5 lead for mallards. Now, I shoot #3 steel.

Common Steel Shot											
Pellet size:	#6	#5	#4	#3	#2	#1	B	BB	BBB	T	
Diameter:	.11	.12	.13	.14	.15	.16	.17	.177	.19	.20	
Per ounce:	315	240	190	155	125	100	85	70	60	50	

Non-toxic waterfowl loads. These 3 ½-inch shells are serious medicine, hitting hard on both ends of the gun.

To improve lethality on geese, some obsolete sizes reappeared, explaining the very large BBB and T listings. The extra punch helps and as you can see from the chart, pellet count can be similar, producing equivalent pattern density. Because individual steel pellets are lighter than lead ones, it takes more to achieve the same overall payload weight. One wrinkle: there is only so much room inside a standard 2 ¾ or 3" shotgun shell. These extra steel pellets occupy more shell space, resulting in lighter maximum payloads or the need for a longer shell. A solution was the new, super-sized 3 ½" 12-gauge hull.

The lighter steel pellets shed velocity more quickly than lead ones, so effective range is somewhat reduced. Patterns also tend to be tighter, due to the harder pellet alloys which are much more resistant to deformation. As a general rule, steel shot requires a whole degree less choke.

Much recent work has gone into alternate alloys of heavier weight than steel. Pellets have been formed from bismuth, or iron/tungsten mixes. Results have been good, but cost is much higher. If you can afford them, great. If not, you can live with steel, recognizing its limitations. It patterns well since the pellets are hard and less likely to deform, but steel loads may damage the barrels of older guns. Regular use may also damage the choke tubes supplied with newer guns. Seek a gunsmith's opinion before firing steel through a questionable barrel or choke combination.

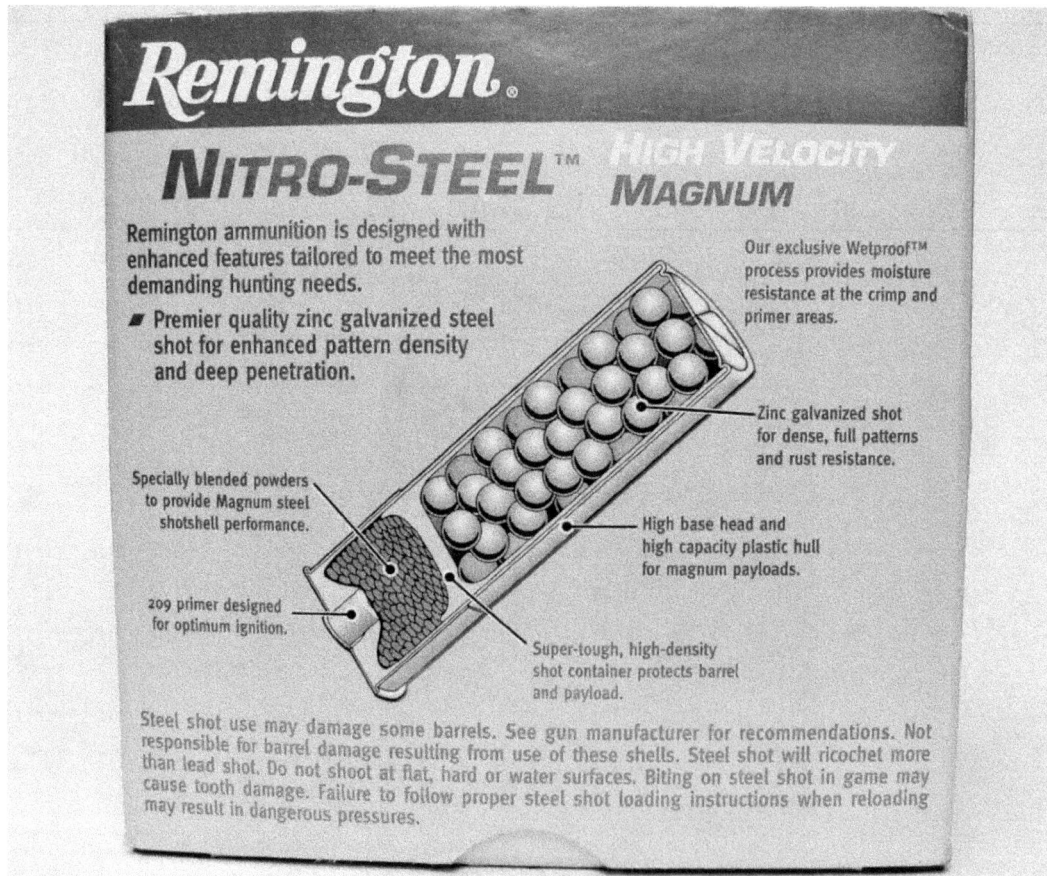

Steel shot; note the warning on the box.

Buckshot. For larger game or self-defense, so called buckshot comes into play. A different grading system determines pellet sizes, which commonly run from smaller 0000-buck to 0-buck. By far, 00-buck is the most popular.

Smaller pellets help improve pattern density in lighter loads.

Common Buckshot							
Pellet size:	**#4**	**#3**	**#2**	**#1**	**#0**	**#00**	**#000**
Diameter:	.24	.25	.27	.30	.32	.33	.36
Per ounce	21	18	14	10	9	8	6

The larger the pellets, the less you can fit in a shell. A 00-buck projectile is about .33 and a standard 2 ¾" 12 Ga. load contains just 9 pellets. Conversely, you can stuff a surprising quantity of small shot in the same vessel. Pattern density will improve, but tiny pellets lose their punch fairly quickly. The reduced energy translates to poor penetration. Like steel shot, one fix involves longer shells capable of holding extra, larger pellets. Some premium 12-gauge loads are available with 12 pellets. By stepping up to a 3" Magnum, 15 00 Buck pellets can be thrown. The 3 ½" Super Magnum holds 18! This major increase comes at the inevitable cost of stiff recoil. I'm still sitting on a small stash of 3 inch, which I'm in no rush to use. A switch to smaller buckshot pellets will increase their

3 ½, 3 & 2 ¾" buckshot loads. The longer shells pack plenty of wallop on both ends!

count, at the expense of less individual pellet energy. The trick with buckshot is to understand its limitations.

More pellets and higher velocities don't necessarily translate to improved patterns; neither do tight chokes (you can skip ahead to catch more information on this). An old law enforcement rule of thumb equates buckshot spread to roughly one inch per yard of travel. In other words, at 18 yards you should see about an 18" pattern. This formula pertains to 00 Buck from a non-choked, cylinder bore barrel. We've done extensive research about the effects of various choke constrictions on patterns. In our experience, a full choke won't perform as well as a modified constriction. Pellet deformation in the tighter choke is a possible explanation. Using the right loads and a modified choke we can extend 18-yard 00-buck performance to about 33 yards. Not bad! One thing to consider: buckshot pellets aren't true bullets. They're spheres with lower mass. A 00 Buck pellet only weighs 54 grains. Velocity is quickly shed and multiple hits are required for effective results. No matter how we slice it, buckshot is a fairly close range proposition.

12 Ga. slug with a mix of 00 pellets, some of which are copper-plated.

Slugs. These are essentially big bullets designed to be fired from shotguns. A plain vanilla, lead 12-gauge slug is .73 and weighs 437 grains. In other words, it's a bullet measuring nearly ¾" in diameter, scaling an ounce! You'll see it listed with a muzzle velocity of around 1600 feet per second (FPS), but it may exit a short barrel at around 1500 FPS. Regardless, its mass ensures good penetration and it'll probably expand upon impact, leaving one hell of a hole! As for accuracy, even common Foster-type lead slugs produce useful groups within reasonable distances. We've fired literally tons of them from rifle-sighted smoothbore barrels. Sighting in at 50 yards, five shot groups will run around 3". They'll have dropped a couple inches at 75 yards, but will still shoot okay. They're often marketed as "rifled slugs." Dissection will reveal a bore-diameter lead bullet with a hollow base. The slug will bear "rifled" ridges on its circumference which, in theory, impart aerodynamic spin after leaving a smoothbore barrel. In actuality, they work more like a badminton birdie, with a similar nose-heavy design. The soft lead composition will safely accommodate varying bore diameters and squeeze through different chokes.

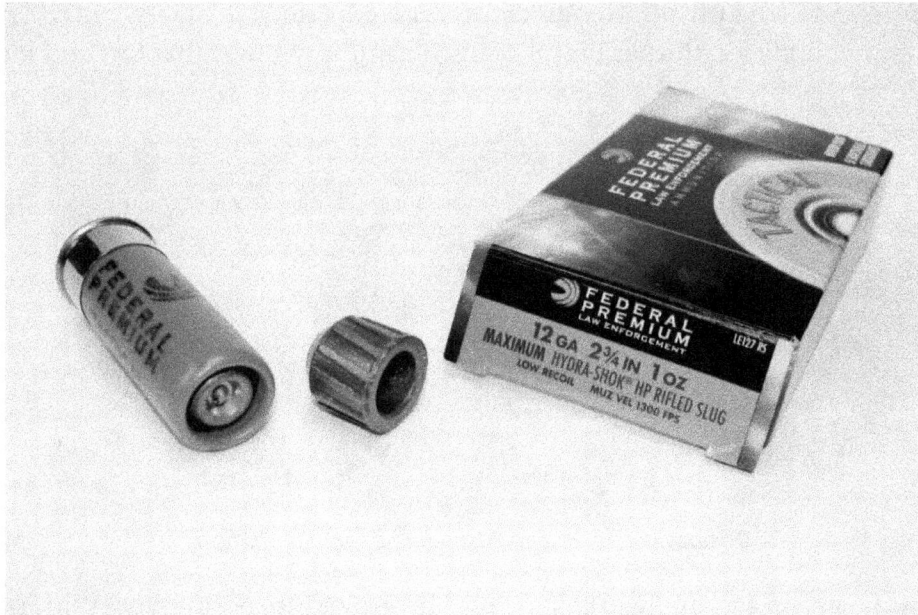

A common Foster-type "rifled slug."

Rifled shotguns can shoot on par with many true rifles if paired with premium slug loads designed for such barrels. We've seen 2" groups at 100 yards from sabot shells, which hit with the oomph of a .45/70. Many shotgun-only deer hunters tag bucks in excess of 150 yards using this technology. Tough, sub-bore bullets are encased in plastic sleeves that separate upon leaving the gun. The sabots impart spin to the fairly aerodynamic bullets, resulting in flatter trajectory.

Most shotgun barrels are smoothbore tubes, designed to expel shot pellets. A problem with rifled barrels involves blown patterns. They're the ultimate solution for improved slug accuracy, but that's really all you can use one for. Plain, soft lead Foster-type slugs will soon smear the lands and grooves with a heavy lead plating so you'll need premium ammunition.

Typical rifled slug accuracy results at 50 yards.

POWER (OR DRAM-EQUIVALENT)

Ever wonder why a shotgun barrel is so much thinner than a rifle? The answer is pressure. Many modern rifle cartridges develop over 50,000 P.S.I. at peak pressure. That's some serious force, requiring lots of good steel for safe containment. The cartridge cases are brass, a durable and elastic metal capable of expanding to contain 25+ tons of peak pressure. Shotguns run at much lower numbers – usually only 25%. You could possibly build one capable of achieving rifle force, but I'd rather watch while somebody else touched it off. Remember, we're expelling relatively heavy projectile mass. Boost pressure and you'll need thicker barrel walls. The gun will weigh a ton and kick you into the next county. Scratch that idea.

The business end of a .30/06 and 12-gauge shotgun.

Instead, a practical system has evolved balancing power against portability. With most peak pressures running below 12,000 P.S.I., we'll still get plenty of recoil from warmer loads but the cartridges can be made from plastic, or even cardboard. The earliest "paper shells" were loaded with black powder, charged in drams. A 12-gauge trap load contained 3 drams of black powder. More potent hunting loads might contain 4 drams. A light load might only have 2 ½ drams. Around 1900, modern smokeless powder caught on and we've been burning it ever since; however, the old-timers needed some frame of reference when matching power to game. That's where "dram equivalent" comes from. A 12-gauge trap load is a warm-ish target shell designed to break clay pigeons out to 40 yards. It'll probably be a 2 ¾" shell, containing 1 ounces of #7 ½ shot, powered by a three3 dram equivalent charge. Muzzle velocity will run somewhere around 1200 FPS. The heaviest 2 ¾" loads use 3 ¾ dram-equivalent charges, for waterfowl use with steel shot. In the stricter sense, drams equate to velocity – the common denominator whether generated by black or smokeless powder.

It begins to make sense once we understand the numbers.

Why do we need this information? This helps us buy shells. Let's look at a few shotgun shells and figure out the numbers. Uncle Harold dies and his widow gives you an old hunting vest. In its pockets you find some shotgun shells with numbers still visible on their bases and sides.

- ➤ **12 Ga.** (indicates a 12-gauge shell) often stamped on the base.
- ➤ **3** (indicates 3 dram equivalent charge) printed on side of shell.
- ➤ **1 ¼** (indicates ounces of shot) printed with the same sequence.
- ➤ **6** (indicates #6 shot pellets) per above.

In this case you're looking at a fairly standard hunting shell suitable for pheasants. Through the 1970s it would have worked for ducks. Then, federal regulations banned lead shot for migratory waterfowl. Steel shot became the standard replacement, followed by other alloys containing bismuth or tungsten. A magnet will tell you whether the shell contains steel shot, and your friendly game warden will probably have one.

SHOTGUN CHAMBERS

You can still find older shotguns with odd length chambers. Often such guns are British side-by-sides with shorter 2 ½" chambers; but during the 20th-century, most gauges evolved to the now-standard 2 ¾" chambering. The 2 ½" .410 is an exception, which only throws ½ ounce of shot.

Magnum shells. Stepping up to 3-inch loads, we can use longer shells to pack in extra pellets. We can also beat the Dickens out of ourselves if we're not careful. I shoot for a living, so recoil is no stranger. I believe many folks are over-gunned. That's fairly common in shotgun land. The 3" Magnums are normally loaded to maximum pressures with very heavy shot payloads. Since for every action there's an equal and opposite reaction, plan on a large gun heading rearward in a hurry. I do shoot a few 3" Magnum 12-gauge shells each season, but use is confined to waterfowl and turkeys - tough birds often engaged at outer shotgun limits. I stick with 2 ¾-inch shells for everything else, and don't find them lacking.

You can safely fire 20-gauge 2 ¾ or 3-inch shells in this well-marked shotgun.

You can measure a shell to determine its length. Allowing for the fired crimp section to open, a loaded shell will be a bit shorter. In other words, a 2 ¾″ hull will run a bit below that measurement – somewhere around 2 ½ inches. Most barrels are stamped with their chambering. European markings are often metric and a 70mm stamping equates to a 2 ¾″ chamber.

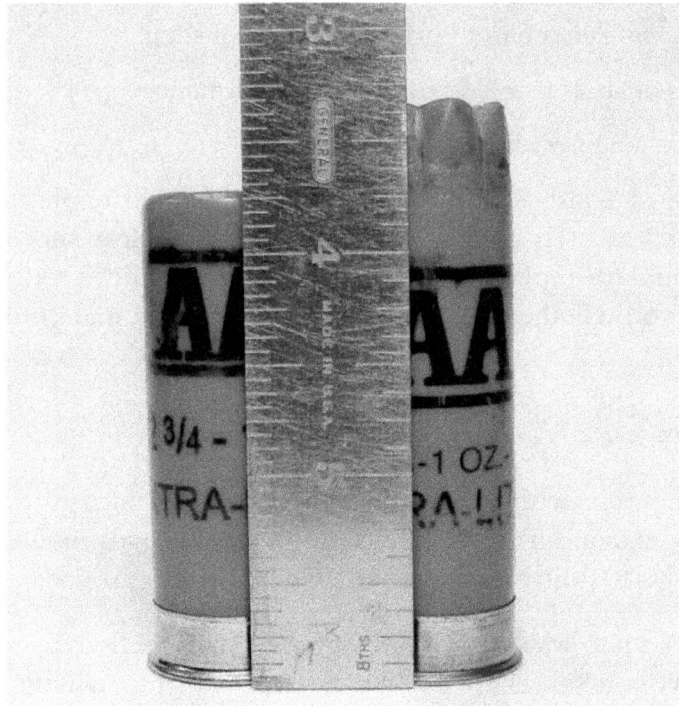

Note the greater length of a fired 2 ¾″ fired shell.

Cautions. Whatever you do, for the love of Pete, don't attempt to fire a shell in a gun for which it is not designed! If you're not sure about the gun, take it to a reputable gunsmith. Firing 3″ shells in a 2 ¾″ chamber is all bad. Along the same idea, you don't want to fire standard 2 ¾″ shells in a shorter 2 ½″ chamber. Compounding the problem, there are still many old shotguns around with Damascus barrels. Usually, these will be double-guns dating from around 1900, or earlier. The steel (which may be iron) displays an intriguing pattern resulting from its manufacture. Metal bands are wrapped, hammered, and welded together around a mandrel, meaning the barrel walls are not a homogenous material. Damascus barrels were developed to handle black powder, which runs at much lower pressure, and is also highly corrosive. Many develop weak spots that will give way with disastrous results when subjected to smokeless pressure. If you have a Damascus-barreled shotgun, treat it as a nice wall hanger.

Don't mix gauges either. God forbid you inadvertently drop a 20-gauge shell in a 12-gauge chamber. The smaller 20 will disappear and stop at its front end. A 12 Ga. shell loaded behind it can be deadly. For those of us maintaining multiple gauges this is a real concern. Other bore obstructions may be less spectacular, but are capable of inflicting serious injury. Some that come to mind include snow, mud, or a wad. If you get a funny sounding report, stop shooting, unload, and ensure nothing is lodged in the barrel.

CHAPTER 4

CHOKES, BARRELS, AND PATTERNS

No in-depth examination of shotgun performance is possible without a look at the various constrictions applied to the business-end of a barrel. A slight reduction in bore diameter near the muzzle can significantly tighten shot patterns – something discovered and refined during the past 150 years. A "pattern" is an expanding distribution of multiple projectiles (shot pellets) which can be recorded for evaluation at a given distance. A standard means to do so often involves firing a shell at a large, blank piece of paper or steel plate. Each pellet strike can then be viewed so dispersion can be evaluated.

Ideally, no large gaps will exist to cause missed targets. A claybird may break from just a single hit but animate targets will probably require several pellet strikes. A nice, uniform pattern is desirable which, when viewed on paper, portrays the appearance of simultaneous strikes. In actuality, a shot charge begins to elongate during flight and at 40 yards it may be several feet long. It will also be a whole lot less dense than it would be at half that distance. As range increases, shot dispersion reduces hit probability. Pellet energy also diminishes and both factors conspire to limit effective range.

Even in the best circumstances, maximum range will probably be 50 yards and, for most of us, 40 yards is the safer bet. Fortunately, many opportunities occur at much less distance, in which case a more rapidly expanding pattern may actually be preferred. We can regulate shot dispersion through muzzle constriction, commonly referred to as "choke."

CHOKES

Until the last 50 years, most common chokes were an integral part of a shotgun barrel, permanently applied to regulate patterns. Then interchangeable, threaded-insert "choke tubes" caught on, providing greatly improved versatility. Now we can switch between tightly constricted "full" chokes or more open increments to optimize patterns at anticipated ranges. A goose hunter reaching up for 50-yard targets will want the tighter chokes. An upland bird hunter in thick cover will need quickly expanding shot patterns and little-to-no choke. The same logic is applied to so-called "riot guns."

A double-gun will typically provide two different constrictions appropriate for its use, and some are still sold with fixed chokes. Such guns, if equipped with a barrel selector switch, or two triggers, permit use of two choke options. Shooters owning older, fixed-choke single-barreled guns, sometimes purchased a differently choked spare barrel. Several firms now offer proprietary choke tube installation services to modernize older barrels.

Fixed-choke barrels in 12 Ga. & .410-bore. Each double-gun offers two constriction options.

The standard formula for determining choke is based on the percent of a shot charge that will fit inside a 30-inch circle at 40 yards. Full choke is the normal benchmark for "good" patterning, with 70% or more of the pellets striking within 30 inches.

Choke	Abbreviation	30" / 40 Yds	Constriction*	Max Range*
Full	F or Full	70% or more	.035	45-50 yards
Improved Modified	IM	65%	.028	40 yards
Modified	Mod or M	60%	.022	35-40 yards
Light Modified	LM	55%	.018	35 yards
Improved Cylinder	I/C	50%	.012	30-35 yards
Skeet	Skt	45%	.005	25-30 yards
Cylinder bore	C or Cyl	40%	N/A	20-25 yards

Constriction in thousands of an inch based on 12-gauge firing lead shot. Figures are approximate.

While a choke's marking may provide a fair estimate of its pattern performance, there is no guarantee. Among other factors, internal barrel and choke diameters can vary to affect results. You'll also see slightly different constrictions from various manufacturers, and some will list diameters in .005" increments. In that case, the above .012 Improved Cylinder constriction will become .010, and Modified will measure .015". These measurements are often used by aftermarket choke tube makers, and things change with the smaller gauges. While a 20-gauge will often use similar constrictions, the 28 and .410 needless.

A look at our more common full choke constriction will reveal that even tight chokes aren't that much smaller than a barrel's inside diameter. The .035 number translates to around $1/32$ inch; but the wall thickening is only half of that, .0175 or $1/64$". Too much reduction will blow patterns and, possibly, the end of a gun off. In fact, choke tube damage can result by firing large pellets (including steel) through the tighter constrictions. A vibrant aftermarket choke tube industry has sprung up to accommodate the latest federally required non-toxic waterfowl loads.

Fixed chokes. Older, fixed choke guns may sustain muzzle damage from steel shot. They're usually choked a bit tighter anyway, so even one deemed safe may eventually develop problems. An old method for a quick determination with a 12-gauge is to see if the muzzle will accept a dime. If it won't, it's probably a full choke. Many older side-by-sides are so choked, the left barrel being full, and the right being modified. The best bet is to seek a competent gunsmith for reliable information.

The same dime won't fit through this 12 Ga. Full Choke muzzle. It's just a rough indicator though.

Assuming you've inherited Grandpa's circa 1955 Remington single barrel Model 870 pump gun, it'll have a fixed choke barrel. If it's marked "full", you have a gun over-choked for most purposes. Here is where we can appreciate the concepts of wide distribution and continuous production. You have a few options:

1. Buy a newer barrel: You can often find used barrels for a fair price. Many older, fixed choke barrels will be cheap, but most will likely be tightly choked. If you spy an I/C barrel grab it! If not, a brand new choke tube barrel is fairly reasonable.

2. Open up the choke: By removal of barrel material, the existing constriction may be decreased. You'll want a competent gunsmith to perform this service, but cost should be less than a new barrel or tubes. I've gone that route with a few guns, enjoying complete satisfaction. The same firms performing custom choke tube installations should be able to provide a quote.

3. Install an aftermarket choke tube system: In the case of a Model 870, it's probably cheaper to hunt down another barrel. For less common or obsolete brands, the muzzle end can be machined and internally threaded, although thin barrels may preclude this option. Cost varies, running from $100 - $300, which may include three choke tubes of standard constrictions plus a wrench. A few firms specializing in this service are Briley, Carlson's, Colonial, and Angleport.

The tight fixed chokes of this double gun were professionally opened up to .005 & .010"; Skeet and Improved Cylinder.

Remington muzzle with interchangeable Rem Choke tubes and wrench.

Interchangeable chokes. Although most interchangeable choke tube systems share similar features, there is no "one size fits all" design. In fact, even individual manufacturers may offer more than one type, and Beretta now lists four different choke tube systems! Regardless, a roughly 2" section of a barrel's muzzle end is bored larger and then threaded to accept corresponding insert tubes. Their tapered, internal diameters are regulated to meet standard pattern classifications, and labeled accordingly.

Within specific types further variations may exist such as flush or extended tubes. Here are just a few of the more common versions sold by major firearms manufacturers:

COMMON CHOKE TUBE SYSTEMS

Beretta - Benelli: The Mobil choke is one of 4 different Beretta types and 2 Benelli versions.

Benelli Crio+: A new, 2nd version, fitting later guns.

Browning Invector Plus: 2nd generation, interchangeable with some newer Winchester guns.

Mossberg – Winchester – Weatherby: Also fits older, 1st generation Browning Invector choked guns.

Mossberg 835 and 935: Standalone system.

Remington: Most common is Rem-choke, not interchangeable with new Pro Bore types.

Ruger: Older guns share Mossberg, Win and Weatherby pattern. New guns use a Briley version.

A diverse collection of extended and flush-fit Rem Chokes.
The more popular brands permit plenty of options.

Flush fitting choke tubes. Many new shotguns are sold with an assortment of flush fitting chokes in common constrictions. Each tube usually has notches in its muzzle end that engage the lugs on an included spanner wrench. After checking to make sure the gun is unloaded, the choke can simply be unscrewed and replaced by another of a different constriction. The choke tubes will bear inscriptions like I/C, Mod, or Full. It's somewhat aggravating to remove a tube in order to read its inscription, so a second marking system is frequently employed. Beretta's tubes have tiny notches cut into their muzzle ends, with more notches indicating less constriction. The big thing to remember is safety. Check the gun and don't look at the muzzle with the action closed!

Extended and flush-fit interchangeable choke tubes.

The four small notches in this flush-fitting Beretta Optima choke tube indicate I/C constriction.

Flush-fit Beretta Mobil Chokes in Full through Skeet constrictions.

Extended chokes. Competition shooters like extended chokes because they're easy to change. Those seen on clay bird courses usually have knurled external grasping surfaces that extend around one inch beyond the muzzle. Sporting clays (which simulate bird hunting) involves different choke degrees depending on target presentations. Shooters manually switch them without a wrench, performing frequent checks so they stay finger-tight. The extended area also provides a handy spot to inscribe each tube's constriction. It may also permit more gradual choke constriction, which improves patterns. Some of the competition models won't handle steel shot or ultra-high velocity loads, meaning they should be kept separate. I like stainless finish, which is a good identification cue.

Three Mobil Choke-thread I/C chokes. The "MR" is a non-toxic mid-range tube.

Clearly marked extended Briley tubes, flanked by standard Mobile Chokes.

Specialty chokes. The above choke installers, and others, also offer a wide array of aftermarket choke tubes to meet specific needs. Many are based on the extended design:

➤ **Ported chokes:** Extended chokes provide room for porting, which supposedly affords two benefits: 1) some are designed to retard, or 2) strip the wad from the emerging shot charge to further improve patterns. Testing we performed with Angleport tubes supports this claim. Recoil reduction is another supposed benefit. There's only one choke I've used that made a noticeable difference, an older 12-gauge Briley Compensator. It's not listed on their website now, but a newer, similar version is.

➤ **Turkey choke:** This is a specialized choice in a category all its own. Turkey hunting ranges may be long and stray pellets in the breast meat are undesirable. These super-tight chokes are designed to focus very dense #4, #5, or #6 shot patterns on the head and neck of a wild turkey. They're too tight for larger pellets, buckshot, or slugs. Mine is listed as having a .060 constriction, nearly double that of a full choke! I have also killed a pile of turkeys with a regular full choke, sticking to distances inside 40 yards.

➤ **Non-toxic shot chokes:** Waterfowl hunters are restricted to non-toxic shot, the most common being steel. Steady use is hard on a choke, so manufacturers offer extended tubes that locate the constricted portion beyond a shotgun's muzzle. If the tube does swell, it won't lodge itself in the barrel. It's a whole lot cheaper to replace a choke. Since chokes behave differently with harder shot, the standard I/C, Mod, and Full constrictions don't mesh. Instead, they may be inscribed "close, mid, or long-range." While you could fire conventional lead shot through them, patterns might vary. A Carlson's MR choke resides in my Beretta AL390 field gun nearly year-round, throwing I/C type patterns with lead #6 pellets. With steel loads, it's roughly the equivalent of Mod. I like dark-finished tubes to help identify them from target versions.

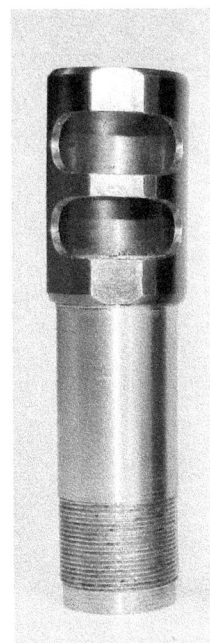

Recoil-reducing 12 Ga. Briley Compensator with IC choke.

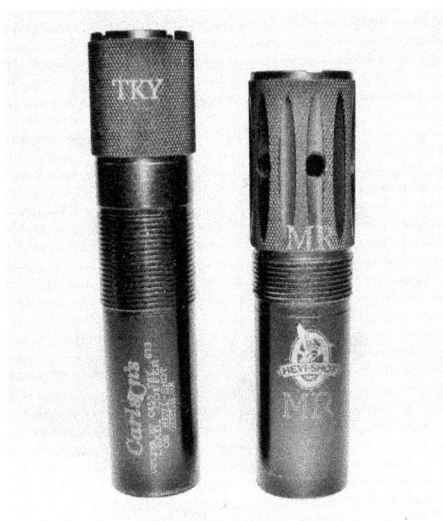

Specialty Beretta turkey & waterfowl chokes. MR stands for mid-range. Note the non-interchangeable Optima-HP & Mobil Choke thread patterns.

➤ **Rifled choke tubes:** Conventional shotgun barrels are smoothbores, lacking the means to impart a stabilizing spin to bullet-like slugs. An expedient solution is a rifled choke tube. Even though its length is short, the spiral lands and grooves do improve accuracy. Sometimes the difference may be slight, but we've seen a few instances where the improvement was dramatic. My son's dot sight equipped Benelli M4 will fire nice, tight clusters with 1 oz. Winchester 2 ¾-ounce plain old Foster-type slugs at 75 yards. His rifled tube is extended, but not by much. We check it periodically so it doesn't fuse to the barrel threads from impact torque and lead. It needs to be brushed out frequently to prevent lead from building, but it works!

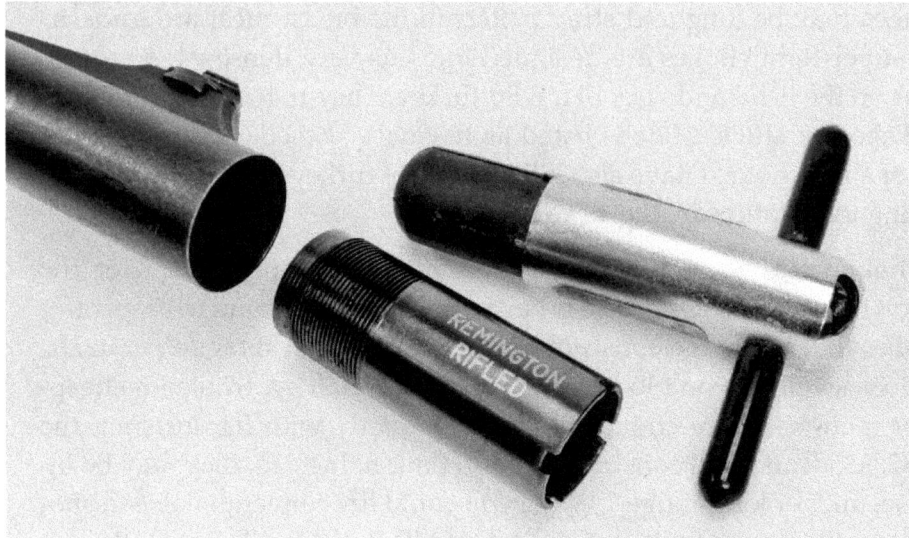

Rifled Rem Choke tube, designed to improve accuracy with slugs.

➢ **Tactical chokes:** A spate of tactical shotguns has appeared of late. They're geared toward the defense market and are often equipped with muzzle devices. Some have a definite cool factor with crenulated faces and lots of ports. The main purpose for their development was breeching. Using special, sintered-metal slugs, the muzzle could be pressed against a hinge to defeat a door. Pressure and metal particles were vented through large ports and the process was exciting. Some operators use fine #9 shot, but they are usually decked out in body armor, helmets, and goggles. The devices in vogue today employ the extended choke principle, meaning they can be easily installed or exchanged. For those who must have one, try shooting it in low light to see if muzzle flash increases.

Tactical breaching choke.

Ammunition and choke effects. A trick that can help expand patterns in fixed choke guns is to shoot cheap, promotional shells. They generally use softer shot that is more prone to deformation. You might see patterns open up by an entire choke level in some cases. Although harder to find and more expensive, "spreader loads" are also available. They're usually small pellet offerings designed for upland birds or clay course games. I use Fiocchi #8 Spreaders in a European, fixed, full choke combination gun, and patterns open up to improved cylinder!

Spreader loads, designed to throw open patterns out of tighter chokes.

Conversely, you can often tighten patterns by firing steel shot. In other words, an improved cylinder choke may throw steel patterns resembling those fired with lead shot from a modified choke. With steel, a modified choke becomes full. Things can become dicey using an actual full choke. Patterns may suffer, and damage may occur; especially with very large pellets.

Choke tube cautions. If you drop a tube, check it for dents. Its lower rim is usually thin and easily damaged. A dent can be hit by pellets during discharge, resulting in gun damage. It's also a good idea to periodically check a tube's tightness. If it becomes too loose, pressure may blow it right out of the muzzle! There is another hazard associated with tighter chokes and certain specialty rounds. Some, like bean bags, travel at low velocity. The wad may lack enough momentum to pass through a tighter choke and, if it lodges in the muzzle, you'll have a bore obstruction capable of causing serious problems. A subsequent shot with a conventional shell will probably rupture the barrel, sending steel shards in all directions. If you ever try such shells, use an open choke and carefully check the bore after each shot. Open the action first and make sure your gun is unloaded!

A collection of specialty ammunition including "less lethal" shells.

If all of this seems a bit daunting, well, it can be! We'll narrow down the list with a few recommended chokes shortly.

BARRELS

Many folks believe shorter barrels will throw giant shot patterns. The myth is further inspired by the legendary "sawed-off shotgun." For starters, federal law restricts ownership of shotguns having barrels shorter than 18 inches. Truthfully, for any dual-role use involving subsistence hunting and self-defense, I wouldn't want a shorter gun. Having used law enforcement guns with 14-inch, cylinder bore barrels, I will say they're handy in vehicles or tight places. They're also really loud! To help keep things simple, we'll cover barrels of more conventional lengths.

Bore and choke relationship. The actual degree of a choke's constriction depends on its size, relative to the barrel's internal diameter. Unfortunately, no precise, universal standard exists among each gauge. While rifle barrels of the same caliber may vary by .001 - .002, shotgun barrels can deviate by three times that figure, or even more. So, an extreme coincidence of choke and barrel tolerances might change choke performance by an entire level. An American 12-gauge barrel is often bored to 0.729", or roughly .73 caliber. We once ran into trouble after installing aftermarket chokes in a series of 18 fixed choke barrels. Although all were from the same well-known manufacturer, a few had much larger bores. Four barrels blew their tubes to parts unknown. Fortunately, the installer blamed himself for not checking each one, and fixed the problem.

Bore diameter. A larger bore, relative to its shot charge will usually throw better patterns. The 16-gauge is widely renowned for firing "square loads." Its modest shot charges have less barrel contact than the same weights in a more popular but smaller diameter 20-gauge bore, which translates to fewer deformed pellets. Similar shot weights can be loaded in a .410, 3" Magnum, or a 2 ¾" 28-gauge shell. The latter's wide shot column encounters less barrel contact, and is respected for performance exceeding what one might expect from such a small barrel.

Understandably then, the latest trend involves larger or "overbored" barrels. In theory, they throw slightly better patterns with less felt recoil. As 12-gauge diameters enlarged upwards to 0.733, or even 0.740", choke tubes needed corresponding redesigns. This explains the sometimes confusing proliferation of choke variations among various firearm manufacturers.

European barrels sometimes run tighter, due in part to use of fiber wads that don't seal pressure as well. My Beretta "Mobilchoke" equipped barrels measure around 0.725"; however, my new Model 1301 has an "Optima HP"

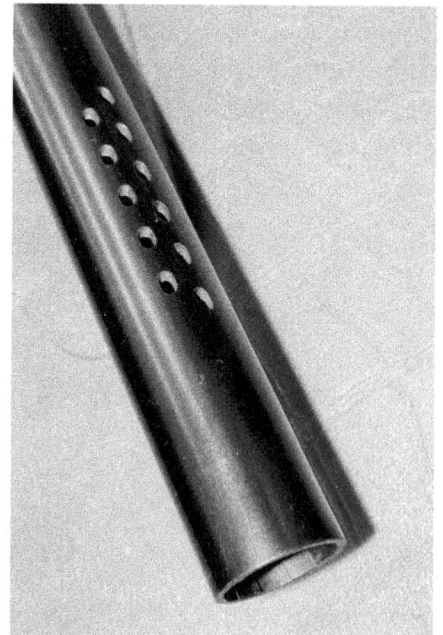

A ported slug barrel. Note the integral rifling, visible in the muzzle.

barrel that measures .0733". Unfortunately, none of my many Mobilchokes fit it. I bought color-coded plastic choke tube cases to help keep them separate.

Barrel modifications. A popular tournament feature is "barrel porting", which supposedly vents off redirected pressure to help reduce recoil. A series of radial holes are cut near the muzzle, which may or may not help. I've owned a few and they looked cool, but I really didn't note much difference. They certainly increase muzzle flash, and also trap fouling during cleaning. "Back boring" is another modification where a rear portion of the barrel is enlarged to gradually convey a shot charge toward the muzzle. Recoil is supposedly reduced as well. Although either may add value to a competitive claybird shooter, they're really not needed for survival-type purposes. In fact, a large diameter bore can sometimes cause problems with wad combinations in cold weather. Stiff wads may not adequately expand to capture pressure, resulting in under-powered "blooper" loads. If you ever get a funny report, quit shooting, unload, and check for a wad lodged in the barrel.

Ventilated ribs. This feature is often seen on sporting guns set up for field or tournament shooting. A "VR" resembles a low bridge with an elevated plane, extending from the gun's receiver to its muzzle. The rib is supported by intermittent pillars, brazed to the barrel. In theory, it provides a seamless sighting system that also diffuses heat. A vent rib isn't essential, but I like one on a bird barrel. For any one-gun owners, it can serve another purpose. A few firms sell aftermarket sighting systems, and Williams offers a neat set of user-installed, clamp-on iron sights. They're a bit fiddly to attach but, once mounted, can convert a bird gun to an adjustable sight slug gun. The red fiber optic front bead stands out well and the sight picture is entirely useable. XS Sights also sells a low profile open sight system, but requires gunsmith installation. The rear sight is an unobtrusive blade, inset in the rib.

Instant slug gun: Beretta 20 Ga. M-391 with vent rib, Williams clamp-on sights and Briley rifled choke tube.

Gauge reducers. These devices are metal inserts shaped like shotgun shells. They are bored through to accept smaller gauge shells, providing the means for some interesting shooting. Skeet shooting is a serious discipline, offering events in four gauges: 12, 20, 28, and .410 bore. Some of the more dedicated shooters shoot one 12-gauge gun, normally an O/U, converting it to smaller gauges with long insert tubes. These adapters are sophisticated designs, closely fitted and balanced to maintain gun handling characteristics - they're also expensive; however, much simpler versions can be purchased, which are really just a hollow metal shell. Surprisingly, they work, just don't expect high volume fire. Two types commonly encountered are branded "Little Skeeters" and "Gauge Mate." Some people report sticky extraction of fired hulls, which is a manual process. Empties can be extracted using a small rod. These adapters probably make the most sense in break action guns, and are sold as pairs for well under $100. Although intended to reduce recoil for tender shooters, one would have the means to fire an alternate gauge in a pinch. Years ago, I owned a Savage "Four Tenner" insert, which was around a foot long and allowed the use of .410 shells in my old two trigger, Stevens 12-gauge double gun. It was fun, and didn't pattern much differently than a true .410 shotgun. The latest versions are only shell length, but accomplish the same thing. A ring of fouling will accumulate ahead of the adapter, but it can be scrubbed out using normal cleaning techniques. Those two barreled guns with inertial triggers may not reset with a very light load, but you could at least fire a shot from its first barrel sequence.

Metallic cartridge converters. Working in a manner similar to gauge reducers, these simple inserts permit use of lower powered handgun cartridges. Some are rifled. Accuracy will be so-so, so useful range will be limited. Check out the stainless steel, rifled "X-Caliber" adapters, which can be purchased as a full set permitting use of common rounds up through .308! Cost as of this writing is $450, but a single caliber unit can be purchased for around $55. Their 9mm and .38 Special units look intriguing, assuming you have a shotgun barrel with some sort of aiming system. They center in the bore using an O-ring index system, and are too long for repeating actions. MCA Sports sells similar inserts in numerous calibers. Lengths are listed as 2 ¾, 10, or 18 inches, with prices running from $35, up to $150. The little 2 ¾" models could be manually fed into a pump or auto, at the expense of performance. The longer versions are fairly quiet, and the 10", $100 unit seems intriguing, but again, a break-action shotgun would be necessary.

SHOT PATTERNS

Considering the variables in tubes and barrels, we can't guarantee that a choke will deliver patterns corresponding with its markings. Even if things are right in one gun, tolerances may produce varying results with another barrel. Shell quality is another important consideration. Cheaper shells generally contain softer shot pellets. The hardening agent, antimony, is expensive, so less of it helps lower cost. Such pellets are more prone to deformation when they pass through a barrel and choke, which can open patterns. Premium shells with harder pellets are tougher and more likely to retain their shape.

Pattern testing. The only real way to know what you've got for a choke is to shoot patterns. Getting serious, you determine the number of pellets in a given load and do the math so you can calculate the percentage of hits inside 30 inches at 40 yards. It's okay to have an "aiming point", which won't necessarily coincide with your patterns. We'll examine that issue shortly.

Meanwhile, you may be thinking, "who cares?" You might if you're looking for range or clout in serious situations!

The assumption is that the pattern is centered on the circle. Reality often dictates otherwise, due to factors like gun fit, recoil, and velocity. One solution involves firing a pattern on a large piece of blank paper, or a painted steel patterning plate (but not with steel shot!). A 30-inch circle is then drawn around the pattern, which is admittedly a bit subjective. Hits are counted to determine the percentage of pellets striking within the circle.

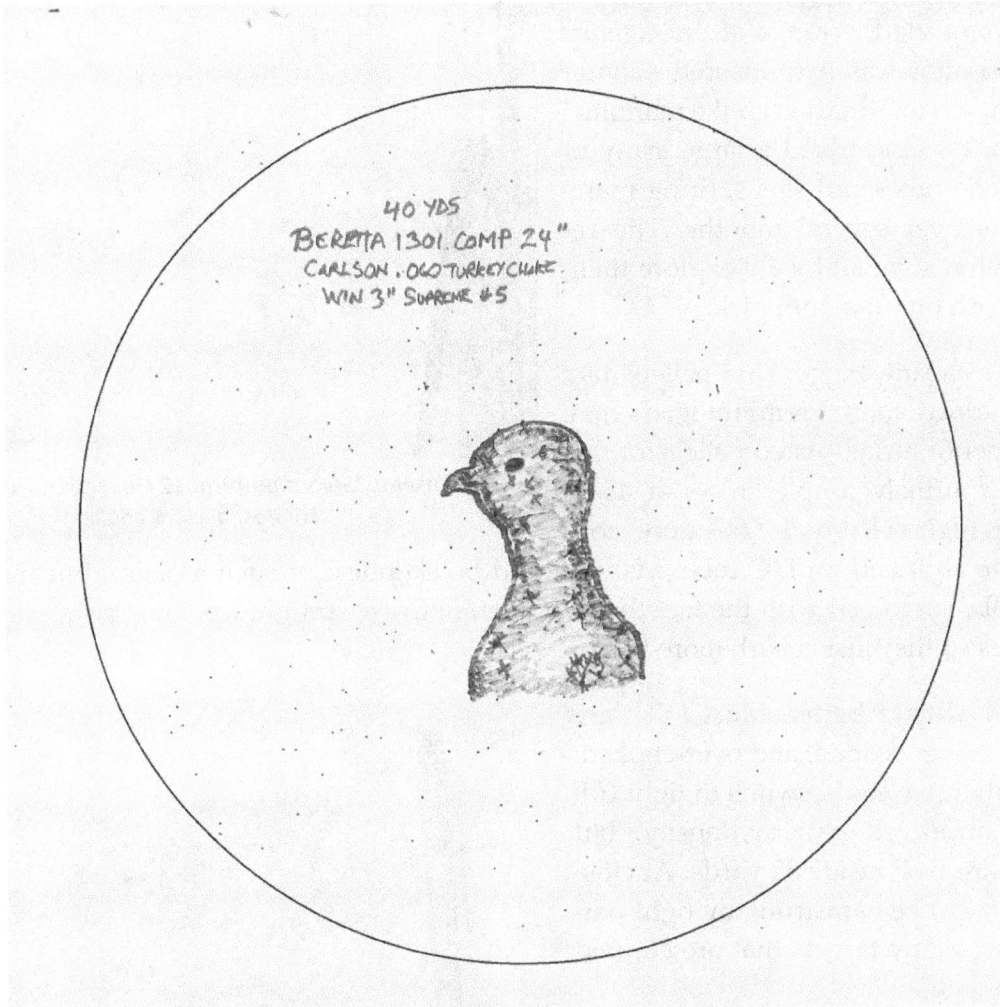

A nice, dense 40-yard pattern fired on a poorly drawn turkey target.

You'll need to know the number of pellets in the load being patterned. An approximate count can be gathered using the charts, but they're not exact. The most reliable method involves dissecting a few shells and tediously counting each individual pellet. Don't be surprised if the numbers vary. In any given shot charge weight, bigger pellets will have a lower count. They may also produce slightly denser patterns due to less deformation. As such, a mid-size load of # 6 shot is a fair basis for evaluating choke performance. From the previous chart, we can see that 225 pellets will comprise an ounce. A 12-gauge, 1 ¼ ounce game load should contain around 281 #6 pellets. If 197 or more fit inside the 30" circle, it's firing full choke patterns of 70%. Don't base this on only one shot. Although more time consuming, reliable data will result from shooting at least five patterns.

Pellet characteristics. We see many buckshot patterns, normally fired from serious fighting shotguns. It's been a real education. You might think the tightest chokes would produce the best results - not necessarily! Several years ago we conducted some research with a goal of extending the range of our agency guns. My co-pilot was a sponsored shooter on a well-known aftermarket choke manufacturer's team. We assembled a large array of different choke tubes and shells, firing many combinations over several months. The results were interesting and we'll explore them more fully with our final gun pick.

Meanwhile, as a rule, bigger shot pellets may require less constriction. Premium loads may contain copper or nickel-plated pellets for this reason. We routinely topple crows at 45-50 yards using high-velocity 1 1/8 ounce cop-

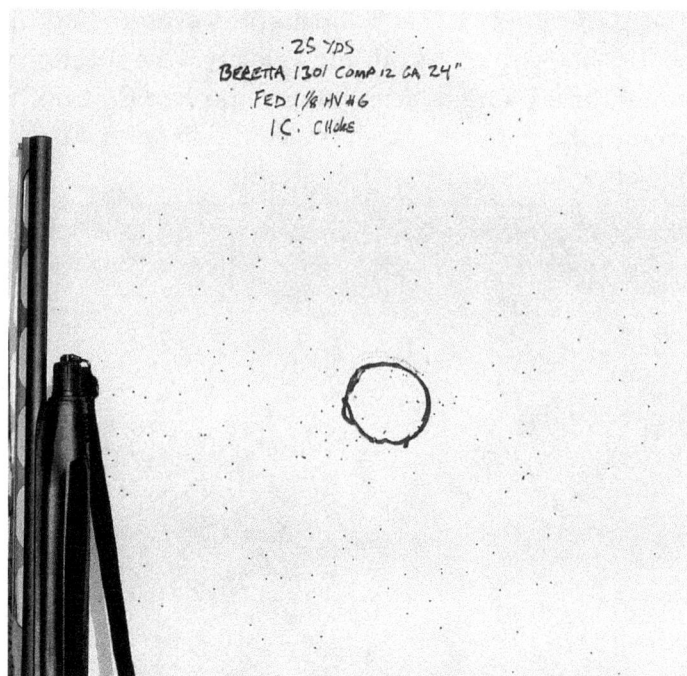

A useful 25-yard pattern: 12 Ga. Federal 1-1/8oz. HV #6s, and I/C choke.

per-plated #6 shot and an I/C tube. Many would be skeptical of such a claim, but the secret lies in good shells, combined with the fact that we're often firing straight up. Face targets expose more square inches so they just absorb more hits.

Tighter isn't always better. Most folks are over-gunned, over-scoped, and over-choked. As for the latter, we could switch to tight full chokes for improved pattern density, but many shots are well inside 35 yards. At closer distances we'd be hamstrung by tight patterns, missing many targets that are clipped by bigger spreads.

Here's another thought: How many people can develop the correct forward allowance (lead) at long range? Screw in a full choke tube and try shooting 25 crossing claybirds at 40 yards to see how you'll do. It's easier said than done. Then try the same thing at 20 yards. You'll still have a challenging target, and your very tight patterns may cause many misses. Switch out to a skeet choke and try those 20-yard targets again. You should see a

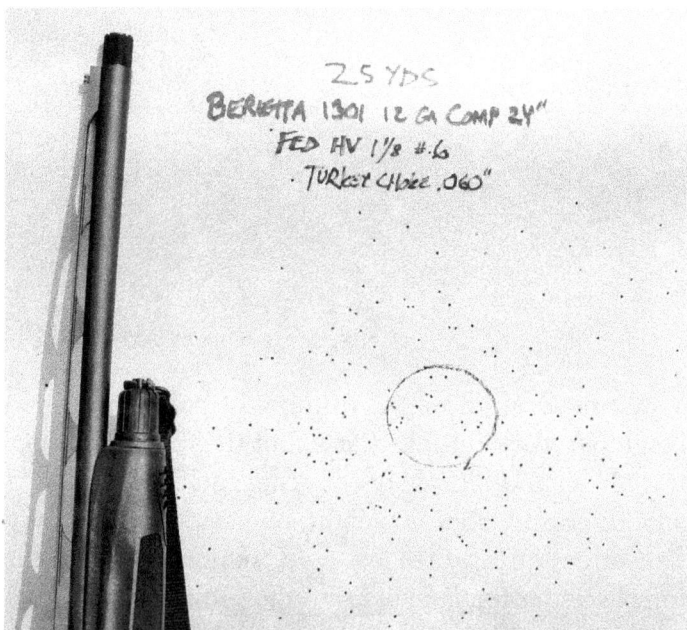

Another 25-yard pattern fired with the same Federal HV load, using a tight turkey choke. Wing-shooting would be difficult with this combination inside 35 yards.

marked improvement. At 40 yards, you won't have a dense enough pattern for reliable hits, so tighter chokes do have their place.

Useful chokes. On the open plains with pheasants, modified may be a great choice. High flying geese may call for all you've got. Turkeys are tough and every hunter is after the densest patterns for neck and head shots. Coyote hunters need a long-range dose of big pellets, focused on a tough and determined adversary. But east of the Mississippi where cover is dense, many shots will happen inside 25 yards. That's where open chokes make much more sense. I'll go with skeet and I/C on a double gun. Using a single barrel, I may start the fall bird season with cylinder bore or skeet, switching to I/C after the leaves thin out.

Many law enforcement or defense shotguns are sold with cylinder bore barrels for a similar reason. When things are close, a wider pattern may be just the ticket. For anyone contemplating use of so-called less lethal, or other low-powered rounds, don't forget that too much choke can trap wads in the muzzle.

Slugs will generally perform best from the most open chokes, but we've had satisfactory results with modified tubes and Foster-type slugs. The soft lead projectile will swage itself through the choke. As mentioned previously, some folks claim improved accuracy with relatively short, rifled choke tubes. In our limited testing we tend to agree. Remove the tube often so it doesn't torque itself permanently into the muzzle threads.

In fact, this is good advice with any removable choke. You'll want to keep an eye on any used with steel shot. The hard pellets can wreak havoc with some, expanding the tube until it's stuck. Extended tubes are a solution to this problem. They also provide a more gradual constriction and give you something to grab hold of. We keep them lubed with good oil, using Triflon or Break-Free. Just remember: Choke tubes are supposed to be removable! Lubricate and tighten them accordingly.

Modified choke results: 00 Buck at 15 yards & slugs fired offhand at 25 yards. The irregular holes are wad strikes.

CHAPTER 5

GUN FIT AND HANDLING

Question: On a bead-sighted bird barrel, what is the rear sight? The answer is your dominant eye. Hopefully, it will be on the same side of your body as your gun. If not, some serious missing is likely. Using the wrong eye will incorrectly reference the muzzle to the target. With the correct eye steering the gun, you'll be looking right over the barrel. We aim sight-equipped slug barrels, which fire a single projectile toward a relatively precise spot. Smoothbore bird barrels on the other hand, are designed to launch an expanding swarm of pellets toward a fast-moving, often airborne target. We must shoot toward a spot where the target and shot will meet, and point the gun, much as we would if pointing at an object with our index finger; our attention is on the object. Staring at the barrel is the kiss of death and results in misses, so we only need a small point of reference (like a bead atop the muzzle) to establish a gun and target reference. It's fast and it works, but only if the gun is pointed where we're looking. As it turns out, many do not.

STOCK DIMENSIONS and PATTERN PLACEMENT

The cheek makes repeatable contact with the stock under proper shooting form, ensuring a consistent eye and muzzle relationship; however, a stock with a comb (top surface) too high for its user will produce the same effect as raising the rear sight on a rifle. It'll shoot high. Conversely, a low comb (known as one with too much drop) may cause under-shooting.

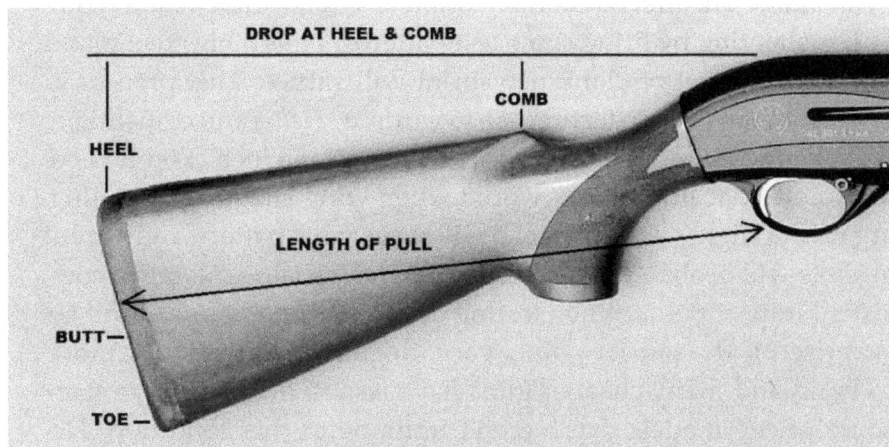

Common stock terms and measurement points.

Cast refers to the left/right component. Length of pull describes the butt-end's distance from the trigger. With compatible dimensions, the stock will position the rear sight (our gun-side eye) for spot-on pointing. Some people spend gobs of money to gain a "fitted" gun. It's an expensive process involving a stock-fitter, try-gun, test patterns, and stock adjustments. The result will be a repeatable gun mount and few excuses for missing. I've gone that route with great results, but have had some success with only a bit of tinkering.

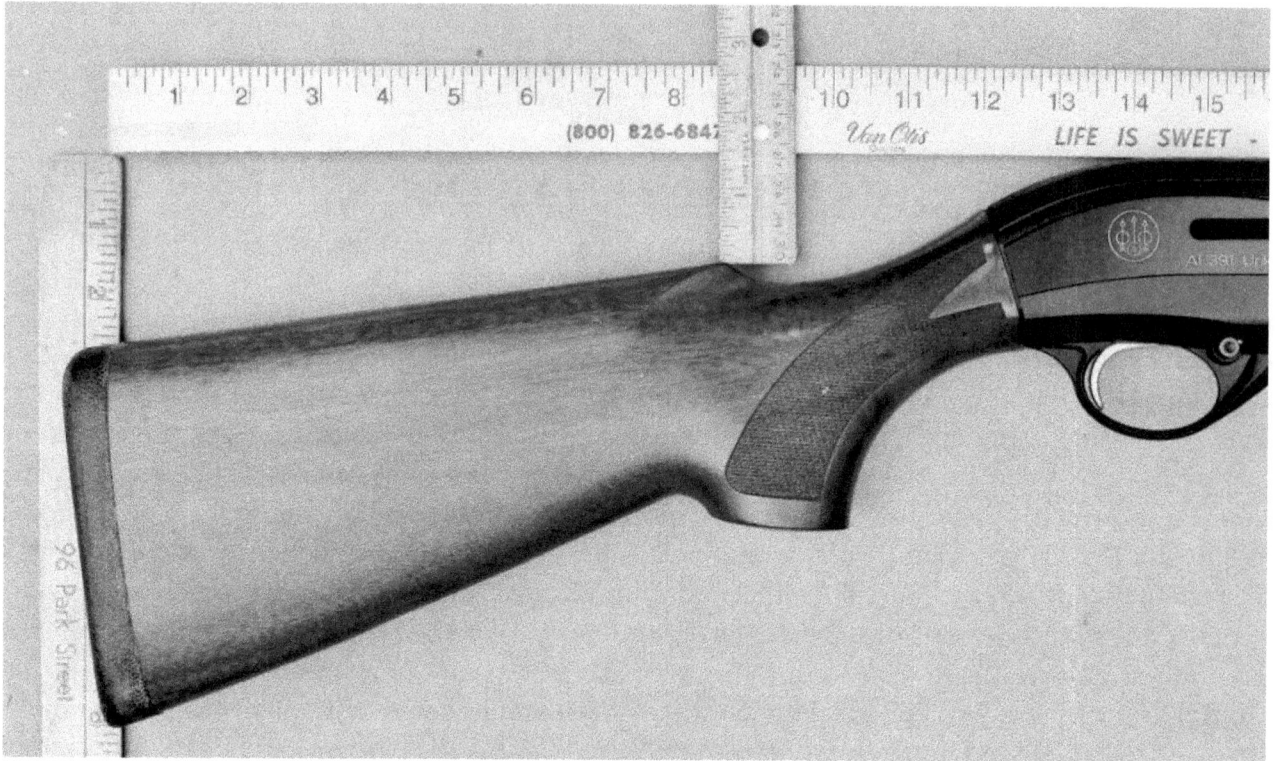

Two measurements determining stock height: drop at comb and drop at heel. These dimensions are fairly standard.

Trial shots. The first recommendation is to determine where your gun is shooting by firing some test patterns. I like a big steel plate, but a large sheet of blank newsprint will suffice. This process is different than pattern testing. At 16 yards, a $1/16''$ comb adjustment will change your pattern's impact by about one inch. You'll need to mount your gun as you would when wing shooting (shooting at birds in flight). A tight choke will focus your pattern and several shots will probably be necessary before you can ascertain a consistent impact point. When testing double barreled guns, don't be surprised if the impact points vary. Regulated barrels cost more to build and many cheaper guns have issues; however, I've seen some pricey models that weren't immune to this condition. No doubt about it, a single barreled gun is a whole lot easier to bring on target.

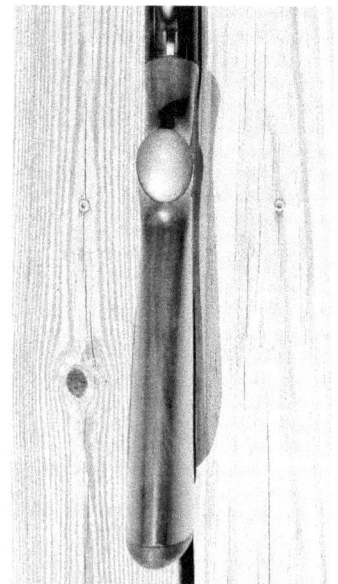

This stock has a bit of "cast-off" to fit a right-handed shooter. A lefty may require a "cast-on" stock.

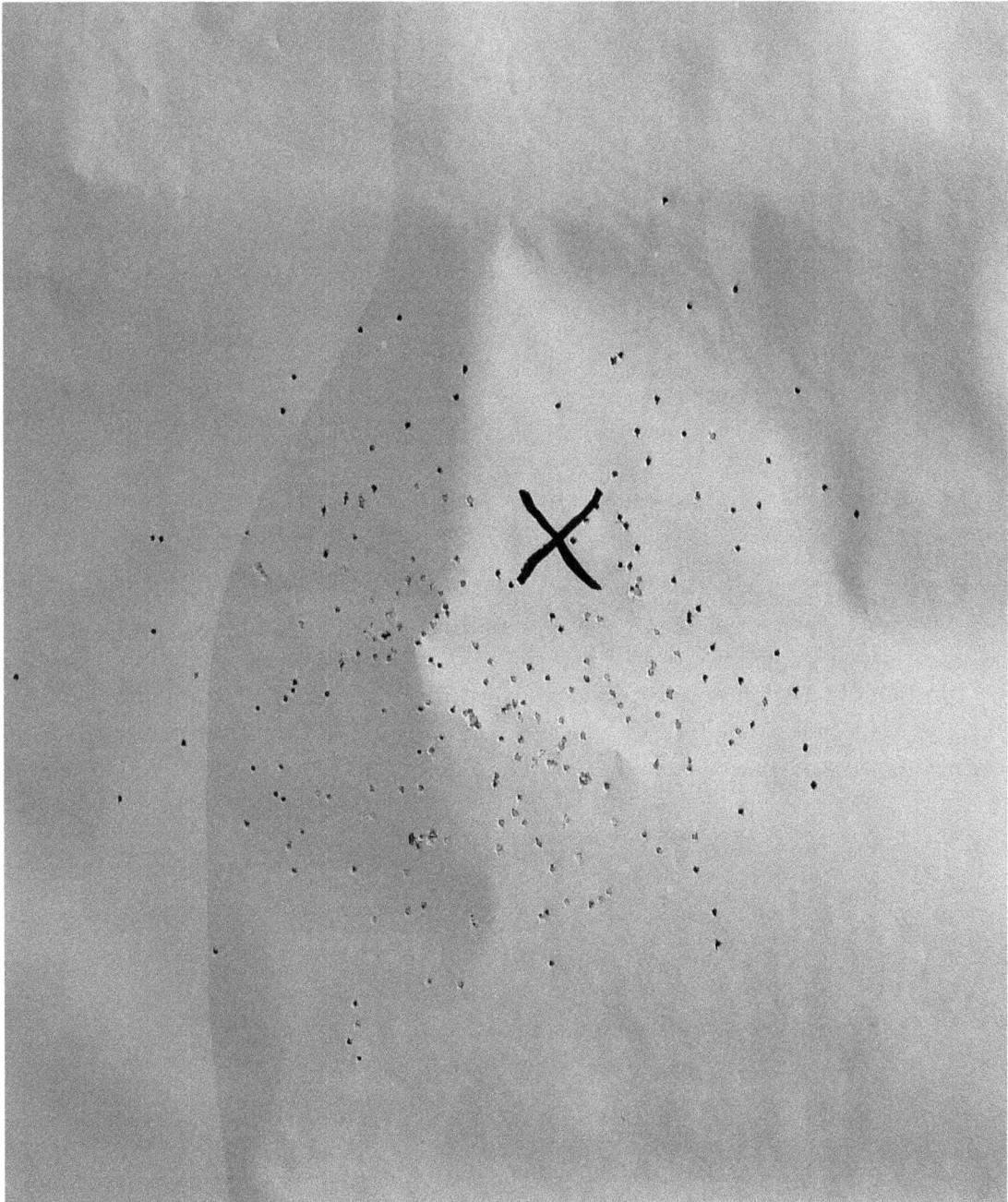

Full choke pattern at 16 yards, in this case striking a bit low.

Adjusting pattern placement. Let's say you've been missing some easy straightaway clay bird targets and suspect something's up. After shooting a few patterns, you notice a low strike tendency. An easy fix may involve increasing your stock's comb height by sticking on an adhesive rubber pad built for this purpose. They're available in different thicknesses and I have a couple guns with them in place. You can jury rig an experimental adjustment to help select the correct height. On the other hand, a stock too high will require more effort. The same applies for cast. Now you know why I like my Berettas, which have spacer adjustments.

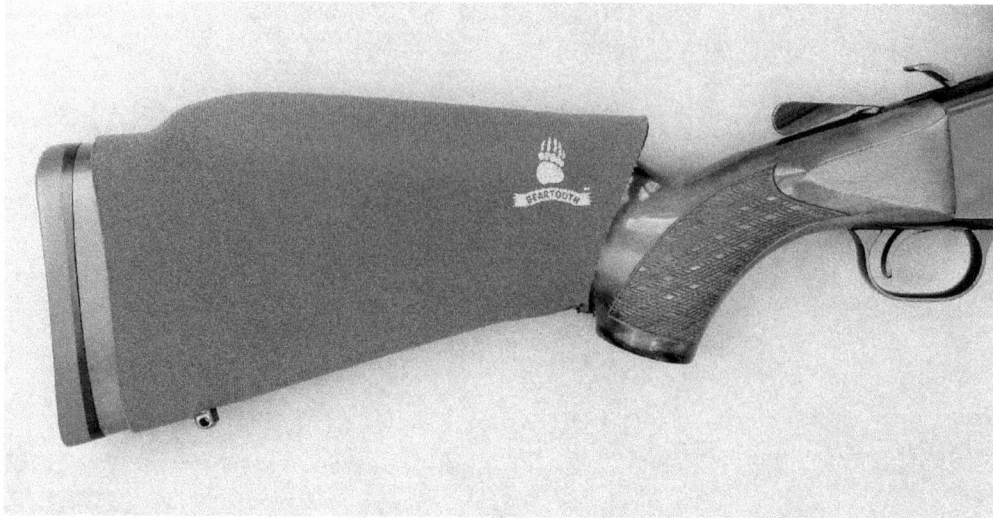

An elastic "Beartooth" comb adjustment sleeve, which uses foam spacers of varying thickness. A small hole was punched to accommodate the sling swivel stud.

I will say this: Too much science can really mess with your mind. My lead instructor is a champion skeet shooter. His present gun, which is worth more than my truck, has never been fired on a pattern plate. For starters, he specified his known length of pull. He has an adjustable comb as well. He sets his gun up until he's consistently crushing targets from all angles and then calls it good. Somewhat similarly, if you already have high confidence with a pet gun you should probably let sleeping dogs lie. This assumes you can comfortably achieve an efficient gun mount, which is tied to your stock's length.

Fine tuning: The cheek-pad decreased drop to raise pattern impacts. LOP was shortened to 14 inches.

Stock dimensions. Stock fit includes a useable length of pull (LOP). "Pull" is the measurement in inches from the trigger to the butt end of a stock. Youth and lady's guns typically have shorter dimensions for good reason. It's discouraging to watch such shooters attempt to climb into an ill-fitting gun. Don't go by the old trigger finger to inside of elbow test. It may work, but other factors can come into play as well. Even a ½ inch adjustment is a huge amount. Stocks can be shortened or spacers may be added. Local skeet clubs or qualified gunsmiths are good resources. A stock too long is very difficult to mount. Conversely, one too short will beat you to death. A good rule of thumb is to go with the longest length of pull you can comfortably manage. Don't forget to factor in layers of clothing, which can make a big difference.

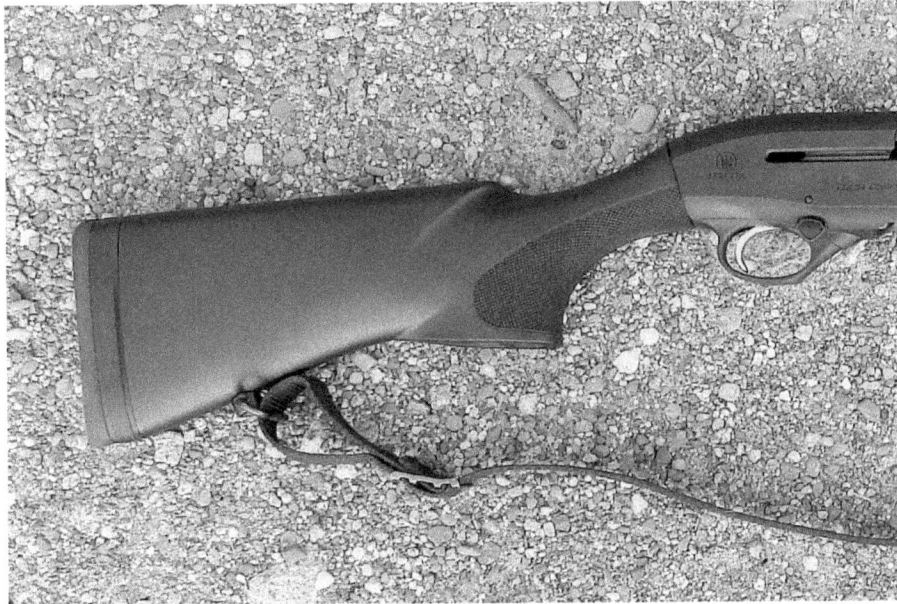

Two spacers included with this Beretta M-1301 were added to increase LOP.

An old trick used to determine proper fit involves holding the butt of the gun vertically in the crook of your shooting arm. In theory, the trigger should be adjacent to the first joint of your index finger. In my case, that translates to around 13 ¾", but I also know that a double gun of that length will probably pound my cheek. Increasing my length of pull to 14 $^1/_8$ " largely solves this problem; so does wearing a thicker shooting vest, a trick that may work for multiple users. When properly mounted, you'll want some space between the ball of your shooting thumb and nose, maybe around the width of a few fingers. This will also prevent a painful punch in the nose during recoil with heavy shells. Besides discomfort, it's pretty hard to make effective follow-up shots with watering eyes!

Most manufacturers strive to build shotguns for a so-called average shooter, who may stand around 70 inches tall and weigh 180 pounds. You'll see differences between manufacturers, with most length-of-pulls measuring somewhere between 14 - 14 ½ inches. If you're built somewhere close to "average", things may be okay. In that case, why suffer through this information? Well, because things can get dicey in a serious shotgun situation.

BARREL LENGTH and PERFORMANCE

Mount your properly fitted, 28" shotgun and point it at a spot some 30 yards away. Now cut off an imaginary 10 inches and note where your new "muzzle" is pointing. It's low. If you bring it up on target your "fitted" gun will be striking high. You've also just discovered a reason why we like 21" barreled, rifle-sighted fighting guns. We've seen tall shooters greatly over-shoot targets when using 18", bead-only barrels. This was in bright light, firing buckshot at stationary targets from only 18 yards. Turn down the lights, add motion, crank up the timer and it gets worse. These things are factored in to an upcoming pick for a do-all shotgun. Don't under-rate gun fit. It can save the day in low light or during split second shooting.

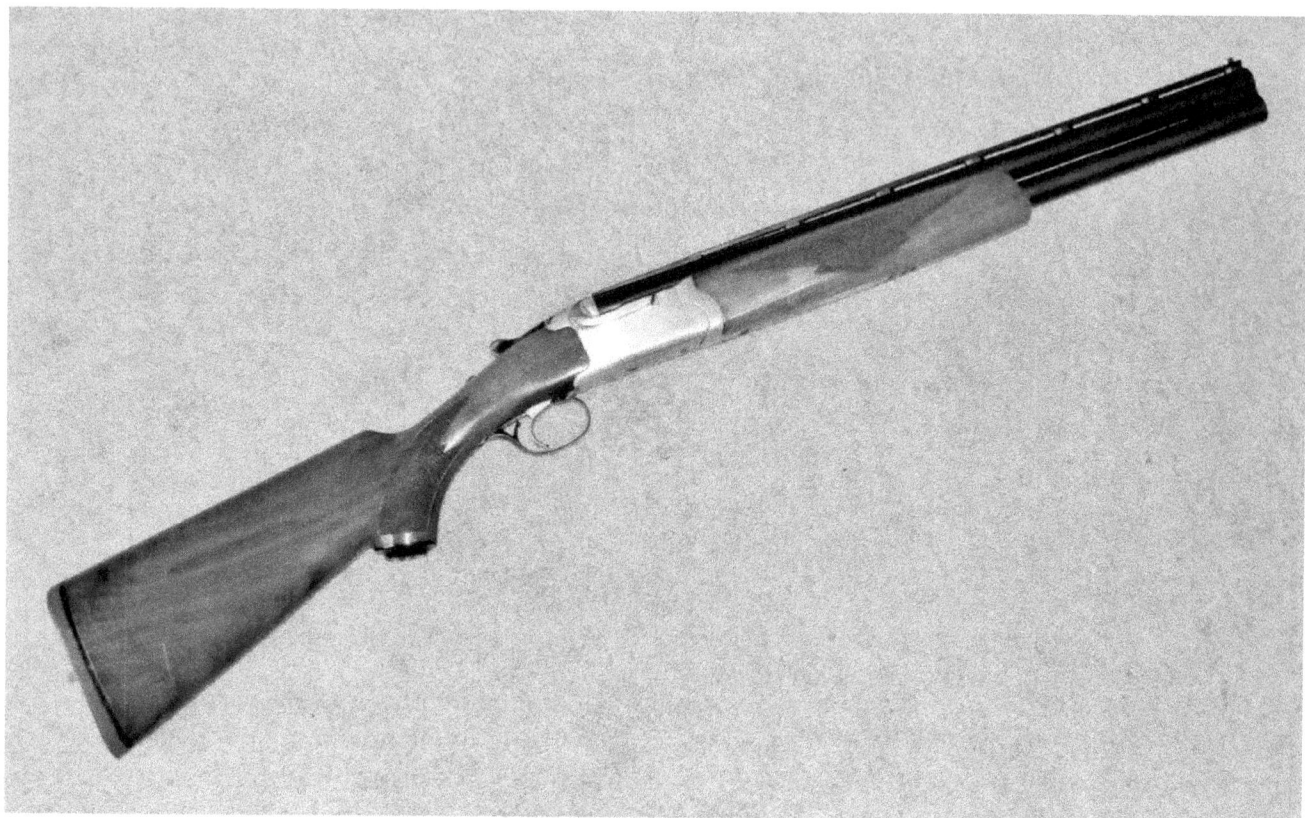

A custom 20-inch Ruger Red Label, which is both short and loud!

My son has a very unusual Ruger Red Label 12-gauge O/U with chopped, 20-inch barrels. It was professionally cut and looks strange to say the least. I always do my best to gain some distance before he fires it. It's loud! Although the stumpy Ruger handles somewhat better than expected, like most short-barreled shotguns, it's still challenging on airborne targets. The British have a term called "preponderance", which relates barrel length to proper handling and a smooth swing. Longer barrels provide the forward momentum necessary to maintain lead on crossing targets. Having played extensively with various lengths in different actions, I concur. There was a time when I'd never consider an upland double gun with a barrel longer than 26 inches. I've completely reversed this opinion, preferring that length as a starting point. The result is a much smoother swing, increased hits, and a better handling gun. The shorter variations feel choppy. Many target guns have 30" or longer barrels for this reason, but for our purposes, and deferring to moderation, something a bit more portable is recommended. One thing to consider is action length. A repeating design has a receiver several inches long. You can add this to your barrel's length, meaning that a 26-inch pump will handle similarly to a 28-inch-plus over/under.

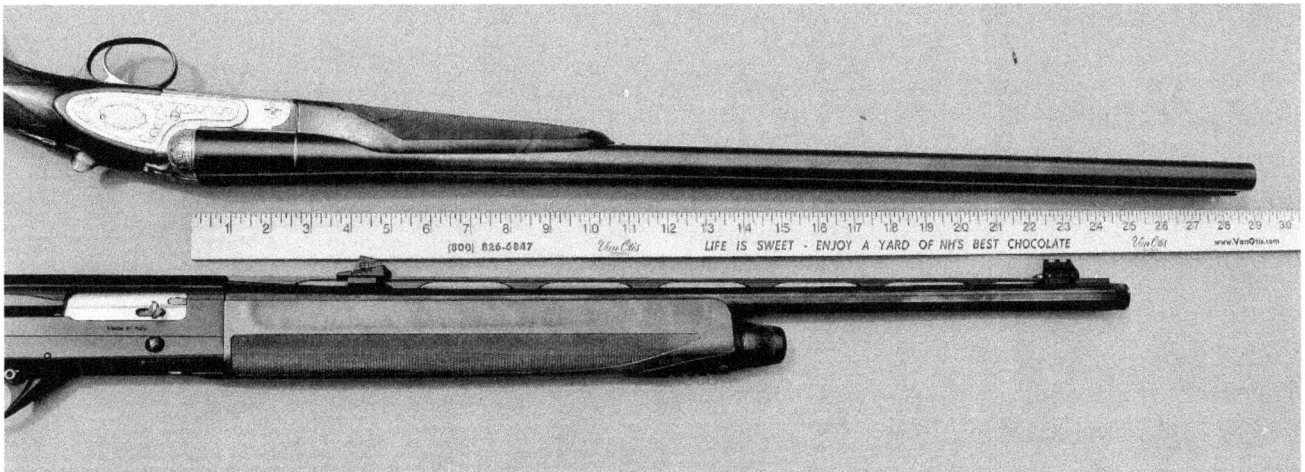

Measuring barrel lengths: The longer SxS is a better overall pointer. The shorter semi-auto has been converted for big game use.

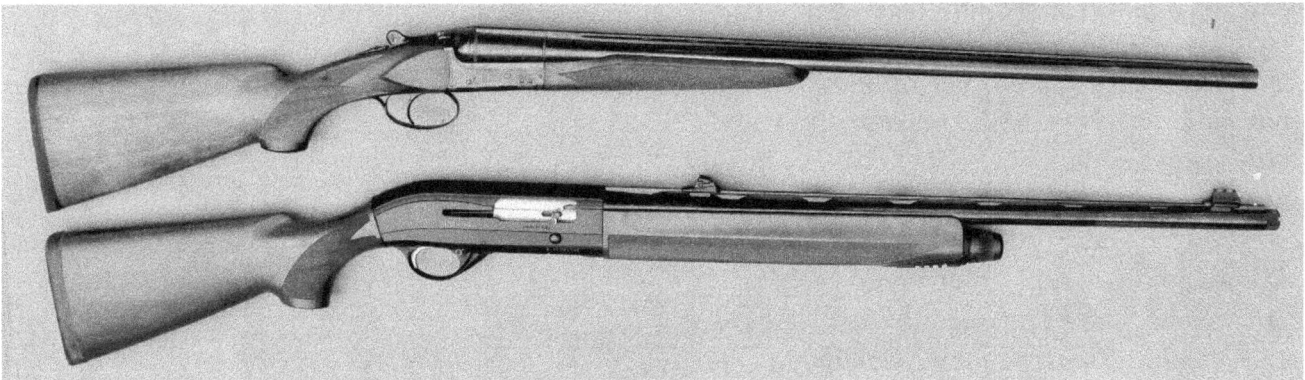

This 26" double-gun is roughly the same overall length as the 24" semi-auto.

If wing shooting is out of the equation, a shorter barrel may suffice. You really won't gain much velocity with a longer one and a shorter barrel carries better when slung over the shoulder. Deer and turkey guns generally have barrels running around 21 inches, and they're often equipped with iron sights or optics. It's no coincidence that fighting guns share similar traits. My Beretta 1301 Competition autoloader has a 24-inch barrel and is a bit of a compromise. Although designed for fast-paced combat courses, it's handy in the woods and isn't a bad all-around gun.

Most people believe short barreled guns will throw wider patterns, but this hasn't been the case in our experience. We recently shot 00 Buck patterns with Federal "Flight Control" loads from 14-inch, custom-built M 870s with fixed Improved Cylinder chokes. The patterns were outstanding and uniform, performance we can quantify after years of experience firing longer barrels. These short-barreled shotguns require Federal BATF permits since 18" is the normal legal minimum.

The best of both worlds can be realized through a switch barrel shotgun. A longer, bird barrel with choke tubes will properly handle all flying targets and a short one with an aiming system can handle everything else.

RECOIL REDUCTION

Some folks buy slip-on recoil pads to help reduce kick. They're a one-piece rubber pad stretched around a buttstock. While they'll work on stocks that are too short, the opposite is often the case. Installation increases length-of-pull, introducing gun mount problems. It's worth spending some extra money for installation of a recoil pad, fitted by a good gunsmith. The stock will need shortening so a manageable length can be maintained. It's a good opportunity to fine-tune stock length, and some of the newer pads do an excellent job of soaking up recoil. As we discussed, adhesive cheek pads can help alleviate cheek pounding to some extent.

Recoil reducers are another option. Some are hydraulic weights that slip inside a buttstock, harnessing inertia to soften kick. Others are just weights. Increasing gun mass helps soak up recoil, but too much causes other problems. Besides winding up with an uncomfortably heavy gun, balance can be affected. Extra weight in a stock can be offset by counterweights up front, and replacement, weighted magazine caps make such fine-tuning easy on popular repeating shotguns. I've used both on my Beretta M-391 target gun to soften recoil and optimize balance. Of course, all the while, gun mass increases.

A well-fitted and effective recoil pad. This one adorns a factory-issued Remington M-870 Wingmaster.

Ideally, we want a lively-feeling gun that balances between the hands, or slightly forward. An old British formula of 1:96 is useful for determining optimum shotgun weight. For each ounce of shot, the gun should weigh 96 ounces. A 12-gauge shotgun firing 1 1/8-ounce target loads should thus weigh 108 ounces, or 6 3/4 pounds. If properly balanced and stocked, the result should be a lively-feeling gun with tolerable recoil.

Firing heavy field loads through the same gun is an entirely different matter. During a high-volume crow control shoot, we were firing 12-gauge, 1 1/8 ounce high-velocity, copper plated #6 shells, rated at 1500 fps. Somewhere into our second box of shells, we both developed splitting headaches. Later comparison indicated recoil similar to 12-gauge 1 oz. rifled slugs! In that case, a heavier gun helps. Better yet, make it a gas-operated semi-auto. A lightweight gun is nice to a point, depending on what you'll be shooting.

High volume crow & headache control tools: gas-operated semi-autos.

Those considering a defensive gun should think about overall weight once accessories are attached. A few that increase heft are receiver-mounted side saddle shell carriers and extended magazine tubes. While these add-ons increase weight by themselves, the greater burden occurs once they're fully topped off with shells. A sling and light pile on more ounces, the cumulative effect being a very heavy gun. We've seen people unable to maintain effective ready positions because they just couldn't support the weight of their gun. Handling quality degrades as well, resulting in a sluggish gun that balances like a truck axle.

Like everything else in life, moderation is the key. The trick is to track down a properly fitted, reliable and reasonably priced shotgun, capable of meeting your needs.

Less than nimble decked out fighting guns. Better eat your Wheaties first!

CHAPTER 6

CHOOSING A SHOTGUN

Let's revisit our selection criteria to help understand the rationale for the shotgun recommendation you'll see in a moment:

1. It must be something with a solid reputation for dependability.

2. It must be easy to operate.

3. Parts must be readily available.

4. Ammunition must be widely available.

5. It must be easy to maintain.

6. It should accommodate practical accessories.

7. It must represent good value.

We frequently get asked advice around a "good" pistol recommendation. The answer isn't always easy and may depend on a number of factors ranging from household demographics through commitment to training. Fortunately, in the case of a shotgun, the choice is a whole lot easier. The gun we'll examine next meets all of our requirements perfectly...

REMINGTON'S MODEL 870 SLIDE-ACTION SHOTGUN

Remington has been "pumping" out a shotgun that remains a best seller since 1950. With over 10 million Model 870 slide-actions in circulation, something must be going right. It's probably safe to say that, even if the iconic M 870 was discontinued today, parts would be available for a very long time.

From its introduction as the Wingmaster over 60 years ago, the venerable 870 has

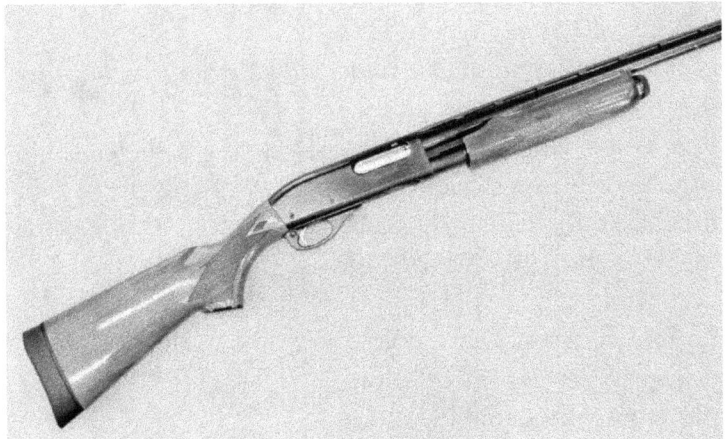

Remington's Model 870 Wingmaster: an American classic pump-gun.

grown a large family tree with offspring ranging from stumpy riot gun versions to long-barreled trap models. It's been offered in all of the common gauges and in several different grades. Since we're focusing on reliable performance at a fair price, attention will concentrate on the more pedestrian examples. The most common version is the Remington Model 870 Express, which has bead-blasted metal and a wood, laminated, or synthetic stock. The Express can also be had as a two-barrel package, usually with a 28" vent rib, choke tube bird barrel and a handy 21", rifle-sighted slug setup. This package would be a darned good choice for all-around utility.

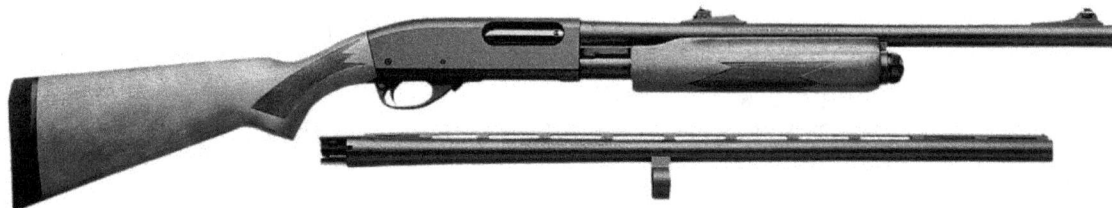

A two-barrel M-870 Express package, capable of covering just about everything.

We inventory an armory full of 870 police models, which are similar to Remington's Express offering. The newest of our 30 guns are 15 years old, but most date to 1991. A satellite group of 8 slightly newer guns has had less, but still substantial use. To be blunt, we've shot the hell out of them using a steady diet of 00 Buckshot, 3 dram trap loads, and slugs. I'm not sure what their lifespan is, but it's certainly not short!

We're running a flight of 11 synthetic stock range guns that have been fired almost daily from May through October for at least seven years. These are the "newer" 15-year-old shotguns. They're beginning to show some wear, but we really haven't had any massive failures. Our older, wood-stocked guns look well-used because they are. They soldier on daily during on-duty missions, where TLC is not priority one. We treat the metal surfaces with Break-Free and clean them when we can. You'll hear stories about the staked shell-stops coming lose, but this has been a very uncommon occurrence limited to maybe three older guns. We've replaced a couple trigger assemblies and a few small parts, but the guns keep running.

An inventory of Remington Model 870s still in use after 20 years.

The trigger assembly can be easily removed by driving out two pins. The barrel comes off after unscrewing the magazine cap. The forend and bolt assembly comes out using just your fingers. Plenty of aftermarket add-ons are available, but some will complicate these steps. A standard 870 with a 26 - 28" barrel handles very well. Go overboard with accessories and you may wind up with something that doesn't.

The 870 Express. Balancing utility against cost, I'd consider a 12-gauge, 3" Express model with a 26 - 28" ventilated-rib, choke tube barrel. I was in a big box store recently (June of 2014) and hit the sporting goods section. As usual, a few Model 870s were on display. The 28" synthetic stock version was priced at $308. An attractive, laminated-stock offering was priced the same.

12-gauge Model 870 Express with 28" ventilated rib bird barrel and camo-finish synthetic stock.

These guns normally come with a removable modified tube. Extra skeet, improved cylinder, and full chokes would expand its usefulness, but installation of the IC tube should cover just about everything. We've experimented with rifled choke tubes, enjoying some success; however, accurate slug placement really requires some sort of aiming system.

An extra rifle-sighted 21" slug barrel solves this problem nicely and can be swapped out in just a few seconds via the 4-shot magazine cap. Using Winchester 2 " 1-ounce Foster slugs, we can shoot 3-shot groups into a 3" circle at 50 yards. A 437-grain .73-caliber bullet leaving the muzzle at around 1500 fps is pretty serious medicine, fully usable to at least 75 yards.

I'd resist the urge to fire screaming magnum loads. They're very expensive, often unnecessary, and likely to induce severe flinching for all but the most seasoned users. We see plenty of experienced shooters with a built-in flinch. It's more common than one might expect and not at all conducive to effective shooting. The giant 3 " super magnums just add fuel to the fire.

The M 870 is available as a left-handed gun with its ejection port and safety reversed, but the standard gun's safety can be switched for left-handed operation. In fact, a right-handed M 870 works well for a lefty, with just this change. You can operate the gun on your left shoulder and throw shells into the loading/ejection port more easily than a right-handed shooter can.

If crusted with ice or dripping with rainwater, you can expect an 870 to work. It'll feed nearly anything and can be disassembled without difficulty. A well-trained pump gun operator can fully exploit this versatile system and is a force to be reckoned with.

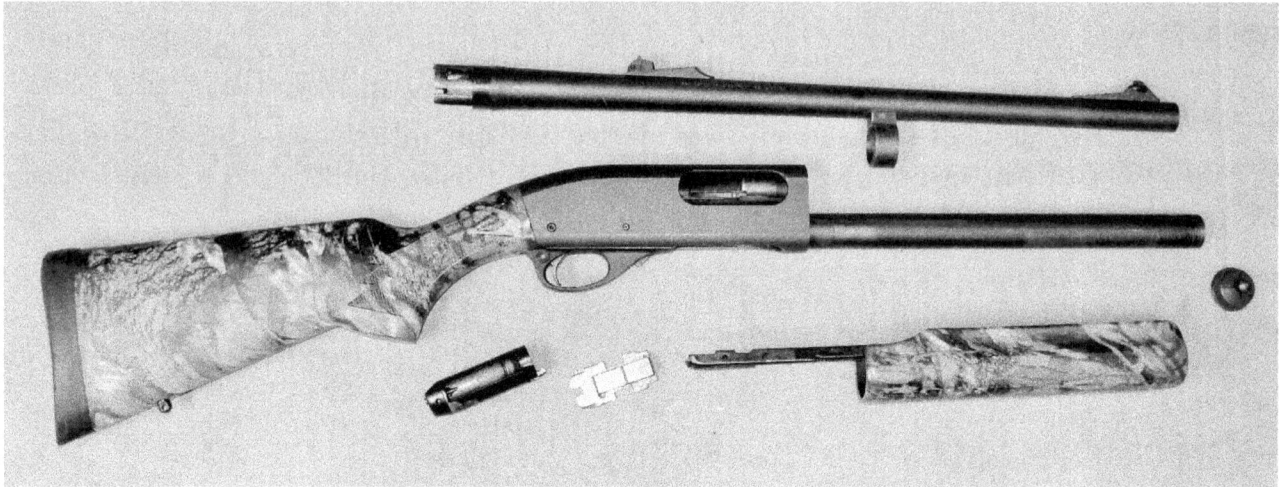

Tool-less field stripping is a snap. As shown here, switching barrels only involves unscrewing the large knurled retaining cap.

SPORTING GUNS VERSUS FIGHTING GUNS

Most of the book thus far has been slanted toward guns that hunt, with additional home defense thrown in. The two-barrel set provides fighting capability via the short barrel option. If you think hunting is an important part of your plan, I'd resist the temptation to over accessorize; for example, extended magazines may seem like a good idea. Trouble is, switching barrels turns into a project. Instead of simply spinning off the magazine cap, tools are necessary. You'll be dealing with clamps and energetic magazine springs that behave like demonic Slinkys intent on popping out during reassembly. Federal game laws also restrict magazine capacity during most waterfowl hunts. The federal limit is just three shells, so a "plug" is included with most repeating shotguns. It is usually nothing more than a removable plastic rod housed within the magazine tube, spaced to accept only two shells. I remove mine for high-volume upland crow shoots, but our state law still restricts magazine capacity to only five shells. In other words, an extended magazine is pointless for hunting.

A true fighting gun is a different animal altogether. It'll hold more shells, weigh more, and probably sport a number of additional features. It may do the job on non-flying targets, but its all-around hunting capabilities will be greatly diminished.

A dedicated defensive gun. For those primarily interested in defensive shotguns, check out the Nighthawk and Scattergun Technologies websites. You'll see some interesting examples of well-equipped Model 870 smoothbores.

We recently sent 10 of our oldest, battle-weary Model 870 police guns to Scattergun Technologies, which is a subsidiary of Wilson Combat. Upon their return, each was refinished with business-like

Parkerized metal surfaces, and new synthetic stocks. We went with short, body-armor compatible stocks, and conventionally designed Davis buttstocks, so our guns would still handle like shotguns. We specified ghost ring receiver sights, matched to Tritium front blades. We kept our Mesa Tactical 6-shell, receiver-mounted side saddle carriers, and located single-point sling plates behind the receivers. The stocks and magazine ends also have QD sling studs. Each gun got a giant head safety button and a high-visibility magazine follower. One big modification required NFA paperwork: a shortened 14-inch barrel with fixed I/C choke. A one-shot magazine tube extension is muzzle-length, providing 5 +1 capacity. The result was a lively-feeling and fast-handling shotgun, good in confined areas.

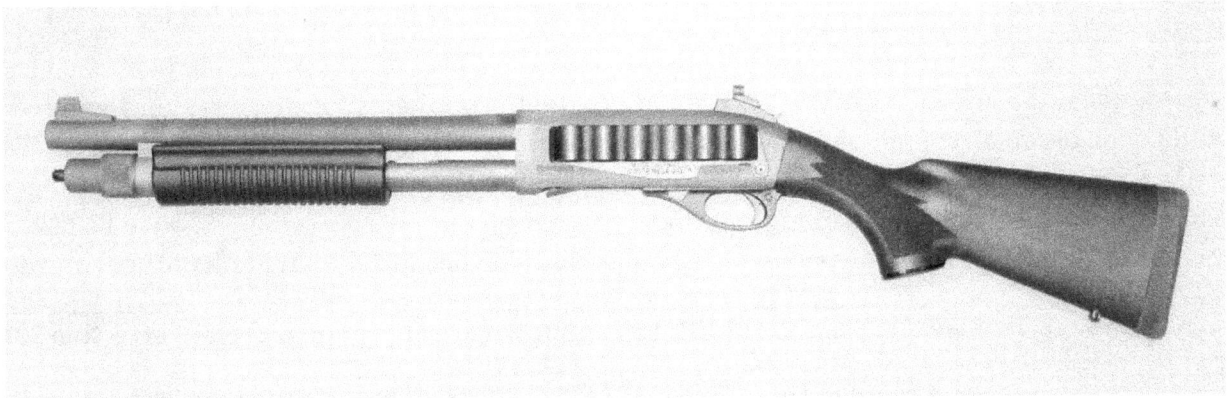

Serious medicine: A customized NFA-type Model 870 with a 14-inch barrel and ghost-ring sights. Note the lack of extra widgets and gadgets.

For less money and no special permits, Remington now offers their 21-inch M 870 Express Tactical. It's a darned good choice right out of the box. You get a black synthetic stock, ghost ring sights, a Picatinny scope base, an extended magazine (6 +1 total), and a cool-looking tactical Rem-Choke device. It's pretty much the whole package. List price is around $600 but street price is less. The extended magazine makes switching barrels more difficult, so those interested in the full sporting gamut would be well-served by this model, buying it as a second gun.

Remington's race-ready Model 870 Express Tactical.

OTHER PUMP GUNS

Not everyone will embrace the trusty Remington Model 870. Fortunately, there are plenty of other choices – too many to cover here. A couple of difficult chores involve identifying which main aftermarket items work with specific models, and nailing down accurate prices.

Prices. The firearms industry is in flux and costs are driven by supply and demand. Some of the more pedestrian models from Remington and Mossberg are also deeply discounted by big box chains. I didn't resort to chicken bones or Ouija boards to nail down prices, but sometimes I came close. The figures shown are a compilation of catalogs, known retailers, and online dealers. There can be a big spread, too. Remington lists the M 870 ET (Express Tactical) for $601, but Gander Mountain advertises it for $499. Cabelas lists several variants under this heading, all for different costs. As for everything shown below, surfing the net may result in a wide price spread; however, odds are the gun will need to be shipped from out-of-state to a local FFL holder. You'll pay shipping plus a transfer fee, which offset savings. Used guns can sometimes be a good bet and *Survival Guns: A Beginner's Guide* has a chapter devoted to the purchasing process.

Descriptions and comments. I try to reserve any in-depth comments for those systems with which I've had considerable experience. While I have some experience with the guns that follow, this list is more of a sampler. Following are a few 12-gauge selections. Some are also sold as smaller 20-gauge versions for similar costs. The Remington Model 870 has already been examined so we'll start with its main competition, the justly famous Mossberg series. The rest are shown in no particular order.

---

12 Ga. PUMP GUN SHOTGUN SAMPLER: Prices shown are approximate retail, effective 2014

Gun	Configuration	bbl	2 ¾"	3"	Ext mag	Side saddle	Retail
Remington M 870 ET	Combat	21"	Y	Y	Y	Y	$500
Remington M 870	Sporting	26"	Y	Y	Y	Y	$320
Mossberg M-500 Flex	Combat	21"	Y	Y	Y	Y	$845
Mossberg M-500	Sporting	26"	Y	Y	Y	Y	$300
Winchester SX-P UD	Combat	18"	Y	Y	Y	Y	$500
Ithaca M-37	Combat	20"	Y	Y	Factory	Y	$880
Ithaca M-37	Sporting	26"	Y	Y	N	Y	$960
Browning BPS	Sporting	26"	Y	Y	Y	N	$700
FN P-12	Combat	18"	Y	Y	Factory	Y	$620
Stevens M-350	Combat	18"	Y	Y	N	N	$275
Benelli Nova	Combat	21"	Y	Y	Y	Y	$450
Benelli Nova	Sporting	26"	Y	Y	Y	Y	$420

Representative models: other configurations and bbl lengths available

---

Mossberg's Model 500 "Persuader."

Mossberg Model 500 and Flex Pump. The Mossbergs are widely used and dependable. They are a good alternate to a Remington Model 870. Mossberg's latest modular "Flex" system permits tool-less adaptation to various configurations. A clever tab permits quick detachable (QD) stock change. The basic M-500 design is more than 50 years old, but several iterations have evolved. As such, complete parts interchangeability is not guaranteed. For example, the similar M-590 uses a different magazine and barrel attachment system. A military M-590A-1 is built to stronger specifications, with a thicker barrel and less plastic parts, including a metal trigger guard assembly. The M-500 trigger housing is plastic, as is the safety. The latter is located in an ideal spot, above and right behind the aluminum receiver. It works equally well for right or left-handed shooters, but the plastic button can break. An aftermarket metal unit solves this problem nicely. A Model 535 variant uses a longer receiver, capable of firing 3 ½" shells. The similar M-835 has an over-bored barrel, not suitable for slugs.

A tricked out Mossberg pump for which accessories are no problem!

In other words, it helps if you understand what you're buying before taking the plunge. There are many Mossberg fans and I won't argue their choice. If you look at the various aftermarket products like magazine extensions and receiver mounted shell holders, Mossbergs will be listed along with Remingtons. The Mossberg product line is vast and interesting! The only real reason it has less coverage here than the Remington M 870 is because I lack in-depth personal experience from which to draw solid conclusions.

Winchester Pumps. The M-1200 appeared in the early 60s as a follow-up to Winchester's famous and well-built Model 12. The 1200 and 1300 pump guns are mentioned here in defer-

The Mossberg's safety location makes lots of sense. Also shown are a side-saddle shell carrier and tactical sling hook-up system.

ence to the many still in circulation. They use a rotary-type bolt that provides a very strong action. The M-1200 was the first popular shotgun to use interchangeable choke tubes, debuting as the trend-setting "Win-Choke" system. Various spin-off models were offered in sporting and defensive iterations. They take down similar to a Remington M 870, but disappeared from production during the 1990s. A 1200 or 1300 is worth snapping up if priced right, and those with one already on hand own a reliable gun.

The latest guns are "Super X Pumps", offered in several variants. They also use a rotary bolt, which cycles very smoothly. Winchester advertises it as the fastest pump gun on the market. Their "SXP Camp/Field Combo" comes with two barrels, one of which is either 26" or 28" vent ribbed. The other is a plain, bead-sighted 18", fixed cylinder bore barrel intended for defensive roles. The VR barrels are machined for Browning "Invector Plus" choke tubes, have 3" chambers, and are internally chrome-plated to help resist corrosion. The "SXP Ultimate Defender", as its name implies, is an 18", purpose-built defensive shotgun with ghost ring sights, rail accessory mounts, and an extended breeching choke. A "SXP Ultimate Marine Defender" ups the ante with a stainless barrel and hard-chromed magazine tube for enhanced weather resistance. Sling-swivel attachment points are provided on these guns, a worthwhile feature. Extra barrels can be purchased as well.

Winchester's SXP "Ultimate Defender", another solid pick.

Ithaca Model 37 Pump. This old model harks back to the 1930s, and was based on a John Browning design, so it *has* to be a good gun! It's different in that it lacks a side receiver port. All loading and ejection occurs through the bottom, a feature that never fails to throw me a curve while attempting to toss in a shell. A local police department in our area uses vintage M-37 "riot guns" and they are still going strong. Like Winchester's old Model 12, they have one interesting trait. If the trigger is held rearward, the gun will fire each time the forend is cycled. It's not my cup of tea and this quirk was changed during the 1970s. Many years ago, Ithaca got ahead of the curve by introducing their short-barreled "Deerslayer" model, which had rifle sights intended for use with slugs. Accuracy was good, in part because this M-37 utilizes a threaded and permanently installed barrel, providing a rigid connection. The M-37 was made in many different configurations, and is a rugged design with a solid steel receiver. Hand fitting increased production costs that drove it from the market at least twice. Some later guns were sold as Model 87s.

The Model 37 has returned once more, now made in Ohio. Among the new models are defensive guns with extended magazines, and rifled Deerslayers set up for scopes. Other new sporting Model 37s have interchangeable barrels, permitting two-barrel combinations. A key point is that the fast "combat load" techniques detailed in the training chapter won't work. Remember, there isn't a loading port to toss shells into. This is more of a technique issue than fault. You'll seldom hear a bad word about the well-respected Model 37!

The tried and true Ithaca Model 37 in sporting dress.

Ithaca's Model 37 "Home Defense" 12-gauge.

Browning BPS Pump. Those comfortable with an Ithaca Model 37 should be right at home with a BPS. It's equally well-made with one distinguishing difference: The safety is located top-side to the tang area, much like a Mossberg. The result is a truly ambidextrous shotgun, and like the M-37, the BPS has a steel receiver. It also loads and ejects through the bottom, offering protection from the elements. Like most other Brownings, the BPS is a higher-end gun, with prices toward the upper end of the pump gun spectrum. All of the popular gauges are offered and numerous configurations can be had. The turkey version has merit for dual-use purposes since it has a set of rib-mounted sights. Another strong point for the BPS is its magazine unloading feature that permits shells to pop free of the magazine tube by simply depressing a shell catch. If ambidextrous function is important, the BPS makes a great choice. Newer guns are machined for "Invector Plus" choke tubes. Because the BPS is sold as a sporting gun, no receiver-mounted shell carrier is available that I am aware of. For me, this would be a non-issue, cancelled out by its excellent safety location. Like the Ithaca M-37, standard combat load techniques are out, but you can still load the magazines of either quickly (as long as the actions are shut).

Another bottom-loading classic, the Browning BPS. Its tang-mounted safety makes lots of sense.

FNH P-12. Originating in Belgium, this firm established an early relationship with John Browning. The Browning connection still exists and FNH now owns Winchester. As such, they know how to build shotguns, enjoying a well-deserved reputation. Many similarities exist across the lines. The FN Pump looks much like a Winchester. Their P-12 uses an aluminum receiver and strong locking system. The 18" barrel is too short for many sporting uses but would make a good home defense choice. It comes standard with an IC Invector tube with others available. A set of iron sights is included and the rear one is nestled in a cantilevered scope base. Because it is attached to the barrel, zero should be maintained before and after disassembly. Capacity is 5 +1 and sling swivel studs come standard. The FN LE is offered with a slightly shorter stock designed for body armor. The 13 ½" pull would also work well for smaller shooters.

FNH P-12 with clever cantilever base and iron sight combination, a dual-use defense and big game package for the price of just one gun.

Stevens Model 350 Pump. Savage has been around forever and the Stevens line is an economy branch of the company. Like the Ithaca M-37 and Browning BPS, it loads and ejects through the bottom. The gun is advertised as an all steel 12-gauge. The "Security Model" has a short 18 ¼" barrel and ghost ring sights. The Model 320 has side ejection and a two-barrel "Combo" set is available for similar cost. Prices are very reasonable. This is one gun on the list that I have no experience with at all, but there are plenty around.

A bottom-loading and affordable defense choice, the Stevens Model 350.

Benelli Nova Pump. Although a more recent design, the Italian-built Nova is fairly popular. The receiver is a steel skeleton, clad in a continuous polymer coating, making the gun extremely weather resistant. Lockup occurs via a rotary bolt head for maximum strength. Several configurations are offered in both 20-gauge and 12, the latter including a 3 ½" version. The "SuperNova" has a juncture between its butt and receiver, which permits different stock choices including a pistol grip version. Benelli's new "Comfortech" stock is also offered, and it really does soften recoil. It uses flexible chevron inserts that act like shock absorbers. Interchangeable cheek pads can be used to fine-tune each shooter's head position; stock shims permit more adjustments. A recoil reducer seems worthwhile when firing stiff loads. A molded-in sling attachment point is another small but handy feature. Accessory barrels are available, including a rifled version for slugs. One interesting feature is a magazine cut-off button, located in the forend. Depressing it permits extraction of a chambered shell while maintaining magazine capacity. This could prove useful if a fast load switch was called for. A transition from 00 Buckshot to a slug is an example. With guns like a Model 870, a shell will feed into the receiver if the bolt is drawn fully rearward. Choke tubes are provided in the Beretta/Benelli pattern. A youth 20-gauge model is available, as are tactical guns. The H2-0 Pump has a shorter nickel-plated barrel with sights for weather resistant use.

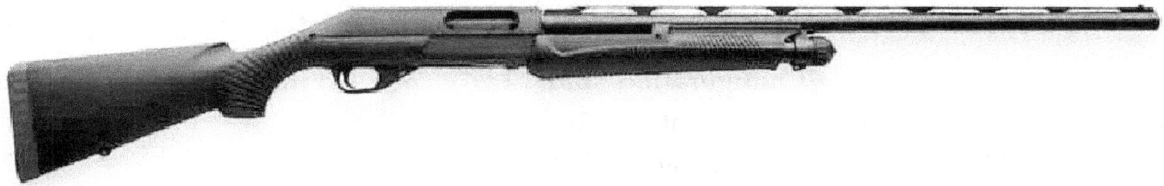

A Benelli Nova pump in sporting configuration.

The cost factor. Without the need for more complicated self-loading mechanical systems, pump gun prices can be priced lower. Exact savings are a variable, but could run as much as 50%. This provides room for accessories ranging from a simple case to an extra barrel.

AUTOLOADERS

We all have personal preferences. Mine lean toward self-loading shotguns. It's not that I'm lazy (maybe a bit), as much as it is a matter of reach and recoil reduction. Many sporting pumps have longer forends that extend rearward to overcome fit problems. When retracted, they also tend to block the bottom port of slide-action receivers. This can interfere with magazine tube unloading, resulting in cycling of each shell through the action. If done carefully, that's fine, but I'd just as soon not do it at all. Some of the autoloading guns share similar magazine clearing issues but most, especially the gas-operated types, are also noticeably softer shooting.

One interesting new shooting game taking shape is 3-gun. This combat-type competition involves fast-paced courses of fire, employing a handgun, rifle, and shotgun. The manufacturers have responded

A rack of tactical-type Benellis, including four Nova pumps equipped with two stock types.

with purpose-built shotguns, normally autoloaders with extended magazines, oversized controls, and enlarged loading ports. Some of this technology extends nicely to defensive roles, resulting in a fast-handling shotgun with generous capacity. They are generally just tricked-out versions of more popular sporting models, meaning some user customization may fulfill this role.

As is the case with pump guns, there are plenty of autoloaders to choose from, and what follows are just a few. A fully decked-out 3-gun competition model will have an absurdly long magazine extension and lots of bells and whistles. It may also have an optical sight and large speedloader bracket. But again, for our purposes, moderation is key. The starting point for such a gun is probably *more* than adequate for most purposes and a not-too-long sporting gun should suffice.

12 Ga. AUTO-LOADING SHOTGUN SAMPLER: Prices shown are approximate retail in 2014

Gun	Type	Configuration	bbl	2¾"	3"	Ext mag	Side saddle	Retail
Remington M-1100	Gas	Combat	21"	Y	N	Y	Y	$1000
Remington M-1187	Gas	Sporting	26"	Y	Y	Y	Y	$700
Remington Versa Max	Gas	Combat	21"	Y	Y+	Y	Y	$1400
Beretta AL300 Outland	Gas	Sporting	26"	Y	Y	N	N	$700
Beretta 1301	Gas	Combat	24"	Y	Y	Y	N	$1200
Benelli M-4	Gas	Combat	21"	Y	Y	Y	Y	$1600
Benelli M-2	Inert	Combat	24"	Y	Y	Y	Y	$1500
Mossberg M-590-JM	Gas	Combat	24"	Y	Y	Y	Y	$800
Winchester SX-3	Gas	Sporting	28"	Y	Y	Y	Y	$1230
Browning Maxus	Gas	Sporting	26"	Y	Y	N	N	$1500
Browning A-5 Stalker	Inert	Sporting	26"	Y	Y	Y	N	$1400
FN SLP MK I	Gas	Combat	22"	Y	Y	Factory	Y	$1040

Representative model: other configurations and bbl lengths available.

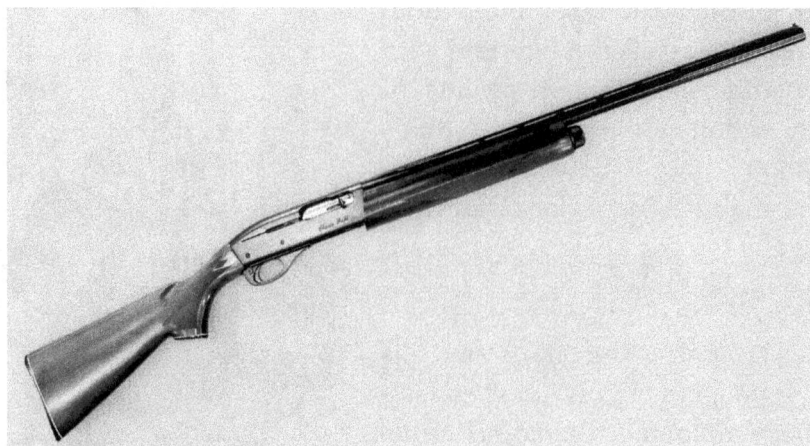

Remington's M-1100 Classic Field Model semi-automatic, chambered for 2 ¾-inch 12-gauge shells.

Remington Model 1100 and 1187 autoloaders. Like the M 870, these guns fit most people. They work with reasonable maintenance and aren't hard to clean. The Model 1100 appeared in 1963, so if something breaks, odds of finding parts are strong. Another plus: It's probably the softest recoiling shotgun on the market. The 1100 was sold to fire either 2 ¾", or 3" Magnum shells, but not both. The more recent 1187 series emerged in 1987 and will feed 2 ¾ or 3" shells without adjustment. A more recent "Super magnum" version, built to handle 3.5" shells in addition to shorter lengths, may be less tolerant of lighter loads. The 1100 series is the parent design but is shell specific. I did manage to wear one out, but it was purchased used and probably digested around 35,000 shells – way more than most people will fire in a lifetime. The 1100 has been built in all of the common gauges, and older guns pre-date interchangeable chokes, but spare barrels can be purchased. They can be changed as easily as those for an 870, but only within their respective model types. In other words, you can't put an 1187 barrel on an 1100. Still, the Model 1100 perseveres, primarily in purpose-built versions. An interesting combat-type Model 1100 is available with features similar to the M 870. For other types, one thing to watch out for is a magazine tube limiter. Remington crimped indentations in some of their magazines near the end of the tube. They reduce the internal diameter to limit shell capacity, which is only an issue if an extension is contemplated. Some people grind or drill them out, and I have swaged out a few to build my own combat-type 1100. If you purchase one, buy spare O-rings to seal the gas system.

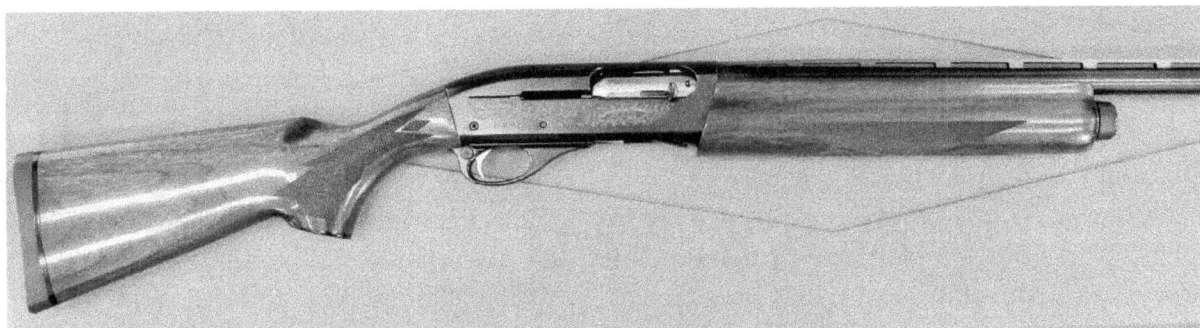

The versatile Remington M-1187 Premier, which feeds 12 Ga. 2 ¾ & 3-inch shells without adjustments.

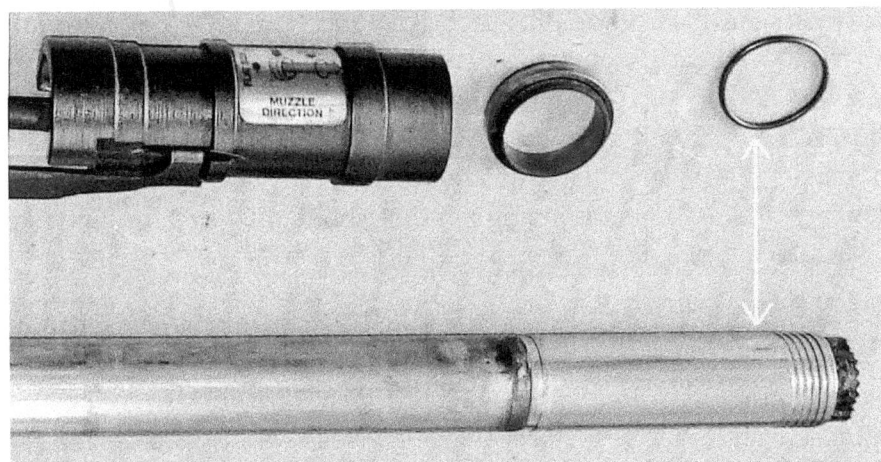

The arrows point to an M-1187's O-ring & magazine tube indent. Note the fouling build-up and directional assembly label. Buy extra O-rings.

Remington Versa Max. This gun is a recent introduction, based on an entirely new gas-operated system. Where most bleed pressure from one or two barrel ports located near the magazine cap, the Versa Max employs a series of radial ports located in its chamber. It will feed everything from light 2 ¾" through 3 ½" shells. Pressure is regulated through the amount of ports available to duct high-pressure gas. The shorter shells use all of the ports while the longer magnum hulls cover many, limiting the volume of potentially damaging pressure. The Versa Max employs a pair of very short pistons located near the chamber, and a rotary bolt head is used instead of the more common single locking lug. It's a strong and simple design that works. Stocks can be adjusted for drop and cast, using a clever mounting plate. Remington continues to expand the line with new, combat-type models, the latest being designed for 3-gun. Beyond its entirely new design, the Versa Max employs a larger-diameter bore that uses proprietary Pro Bore choke tubes. Cost is considerably higher than the 1100-series guns, and because it is a new design, parts could be harder to find.

Remington's innovative Versa Max "Tactical" model, complete with Picatinny rails.

Beretta autoloaders. I own four, most of which have been in continuous use for around 10 years. No doubt all will outlive me. They're reliable and nicely made. Beretta's guns are always a work in progress, so you'll see several gas system variations as their line evolves. My AL 390 is a bit different than the newer AL 391, which has fewer loose parts. Still, I find the AL 390 easier to clean. Many competitive clay bird shooters use AL 391s because they are reliable and handle well. Another plus is the stock adjustment feature, which uses eccentric shims to change drop and cast. With a bit of tinkering, most people can position their stock for centered patterns. My reliable (and attractive) wood-stocked 391s are clay bird guns and have fired *many* shells. My trusty synthetic 390 has seen hard use in woods and waters, never choking during tough conditions. It's the only gun I ever bought at Wal-Mart, and cost $530 – an exceptional value at the time. The 3901 is a slightly upgraded version purported to be another solid choice. I recently handled a new synthetic-stocked AL-300 Outlander, which was bargain-priced at $650. It's a hybrid AL-391/400 and has been getting great reviews. Most of the 390, 3901, and new 300s use Beretta Mobilchokes. Other recent models include the Extrema and 400 Explore, which are different designs, using another choke tube system. The Explore is available as a 2 ¾ / 3", or separate and heavier 3 ½" gun.

Just to add a bit of confusion, a Model 1301 Competition 12-gauge shares the Extrema's 3 ½" receiver, but is limited to shorter 3" shells. It is marketed for combat-type action shoots with its shorter barrel and stock. Weight is minimal resulting in a very lively gun. The longer receiver has larger ports to support fast loading. The bolt handle, safety, and bolt release are oversized for rapid engagement. It is available with a 21" or 24" barrel. I chose the longer version, which fires 2 ¾" or 3" Magnums, making it a viable combat, field, or turkey gun. One kicker: It's machined for Optima HP choke tubes, which won't interchange with the older and more common Mobilchokes. Furthermore, the 1301 only comes with one tube in I/C constriction. I called Carlson's before the gun arrived and soon had an extended turkey tube on hand. The following week this combination effectively tipped over two nice gobblers, using Winchester Supreme XX 3" #5 shells. Most 1301s will be seen on 3-gun competition courses, fitted with extended magazines. Having banged claybirds with the 24" version, I find it surprisingly shootable and it's quickly becoming my go-to gun. I haven't found any receiver-mounted shell carriers for these guns, which doesn't break my heart. They handle well as is and I'd hate to spoil them with extra baggage.

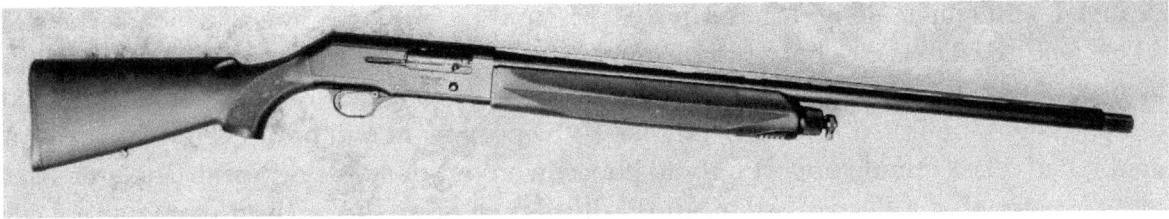

The simple and reliable Beretta AL390, which feeds everything from light 2 ¾" 12-gauge shells or stout 3" Magnums. This one has been going strong for 10+ years, but has since evolved.

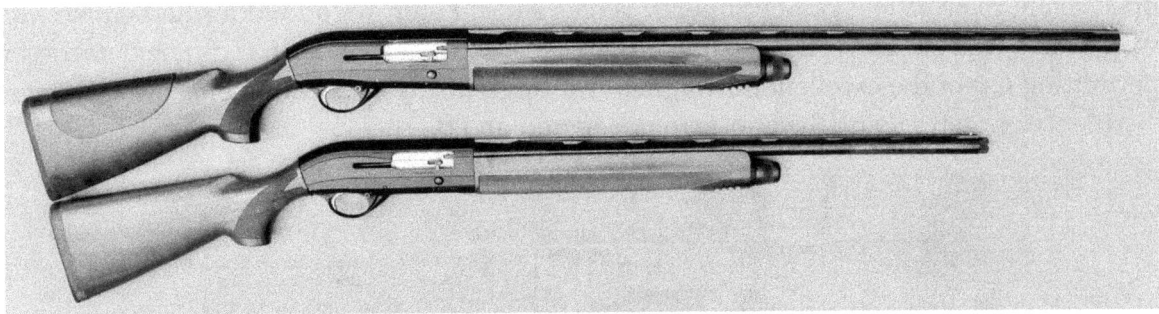

The next evolution: Beretta AL391s in 12 & 20 Ga. The top gun is set up for sporting clays with a 30" barrel. The smaller 20 Ga. is a youth & ladies' model with a 24" barrel. Either gun will handle 2 ¾ or 3" shells.

Further development: An AL-400 based Beretta M-1301 Competition. It's a 3-gun version with oversize controls and entirely different internals. It's a light and handy choice for other uses, too.

Benelli M-4 Autoloader. Developed for military use, this gun is gas-operated. Many consider it among the best. The few we've tested reliably fed a wide variety of loads. An M-4 is not cheap, but it does have a number of good features. My son shoots his way more than I would without maintenance and it still chugs along. It has a pistol grip and desert camo finish, which certainly turns heads. One feature I like is the choke tube arrangement, which accepts Beretta's commonly available Mobilchokes. Although designed as a fighting gun, it works well on turkeys and makes a good deer slayer. The receiver-mounted Picatinny rail has a QD peep sight and optics can be easily attached. He recently had a last-minute opportunity to jump on an out-of-state deer hunt. The area was shotgun-only. We mounted a Burris FastFire III (FF-III) dot sight on the rail and screwed in a cylinder bore Beretta Mobilchoke from my collection. Setting up at 50 yards, we first zeroed using the peep sight. Then we dropped on the FF-III and cinched it with a quarter. Very useable 3-inch groups developed using plain WW 1-ounce Super-X Foster slugs. Later, we added a rifled choke tube that produced even better accuracy. If the battery-powered dot sight does quit, it can be detached with a coin, permitting use of the excellent iron sights. Although it's by no means a bird gun, I've seen him hammer plenty of crows using the ghost ring peep and an I/C choke.

Benelli's M-4 gas gun. Although intended for defense, this one has piled up a bunch of turkeys. The good sights and special turkey choke don't hurt.

The same M-4, equipped with a Burris Fast-Fire dot sight and rifled choke tube. Bench-rested groups resulted in surprisingly good accuracy with common Foster slugs.

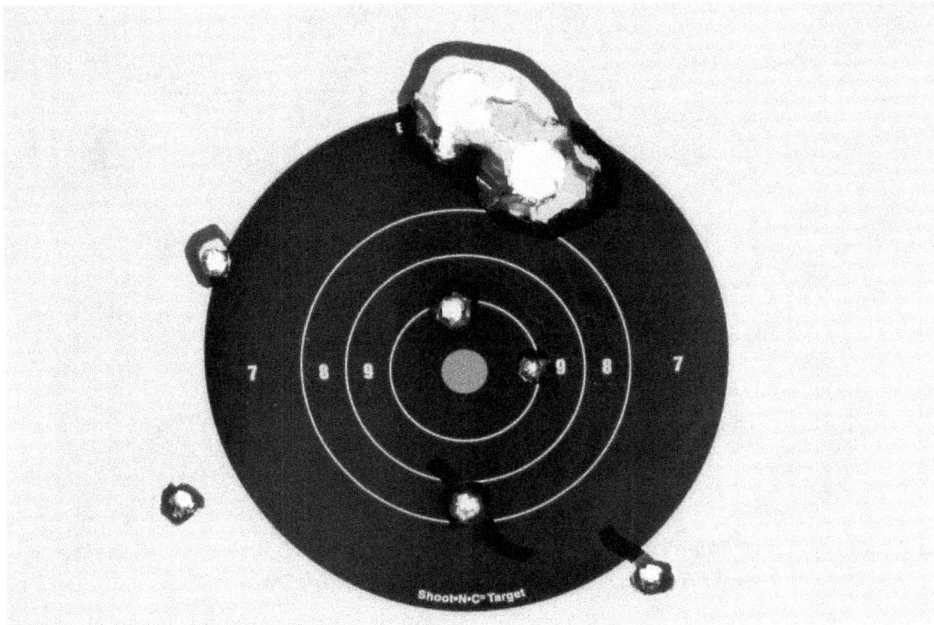

M-4 / dot-sight accuracy: three Winchester 12 Ga. 1-ounce "Super-X" slugs, fired from 65 yards through a rifled Briley "Diffusion" choke tube (the extra holes are from previous rifle shooting).

Benelli Inertia Driven autoloaders. As its name implies, inertia drives the bolt; in other words, there is no gas system. The associated crud is gone and, with fewer parts inside the forend, a trimmer gun is possible. Waterfowlers fire heavy loads in really bad weather, and downpours can be problematic for some gas-operated shotguns. Not so for the Benellis, explaining their popularity with diehard duck hunters. The bolt carrier's mass is harnessed to disengage a rotary bolt and cycle the action, using only a few robust parts. Without the staged recoil impulse of a gas system, recoil is more pronounced, but Benelli has designed an innovative stock that absorbs some of that punch. It's the same "Comfortech" design offered for the SuperNova Pump, and comes with shims for individual fitting.

My instructor colleague, Mike, shoots a 3.5-inch, camo-clad "Super Black Eagle" model with good results. He likes the interchangeable comb feature and sprung for a higher one to nail spot-on patterns. Mike notes that positive function depends on a solid gun mount. In fact, the more complicated gas-powered Benelli M-4 was an answer to inertial impediments caused by mounted accessories such as lights and infrared (IR) illuminators. Cleaning an inertial Benelli is a snap due to its simple design. The gun handles nicely and the chokes sold with some models interchange with Mobilchokes. Later versions use Benelli Crios, but come with a nice assortment of tubes. Among their numerous models, the latest is the high-end "Ethos", which is priced accordingly. No Benelli is cheap, but the more pedestrian M-2 is a sweet-handling gun that feeds 2 ¾" or 3" shells. A 3 dram load is the recommended minimum, but my son's sporting M-2 even feeds Winchester AA 1-ounce target loads! The simple and reliable inertial guns are a popular choice among the 3-gun crowd for good reason. Extended magazines and controls are standard fare, and such configurations also extend to defensive roles. The most unusual model is the M-3, set up for defensive use as a hybrid semi-auto / pump! It's a clever idea that overcomes one pitfall of any semi-auto design, the ability to function reliably with a full range of ammunition like low-powered, less-lethal rounds.

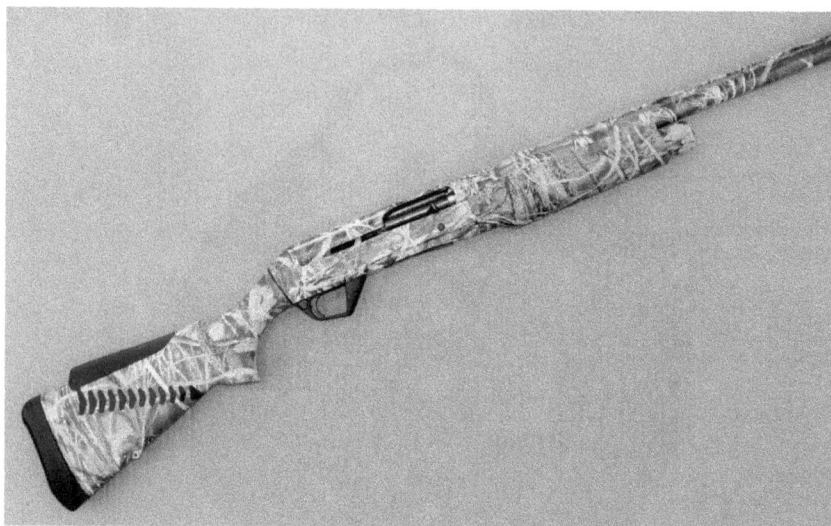

*Benelli "Black Eagle" in waterfowl colors. It will handle 2 ¾; 3; or 3 ½" shells.
The stock helps soften recoil with stiffer loads.*

A Benelli M-2, factory-equipped for defensive roles. Our cadre loved it!

Mossberg. Although symbolic with the well-established M-500 Pump, this firm is no stranger to self-loading shotguns. Among them are their 3 ½" Model 935 series, and 3-inch M-930 guns. Some of the M-930s are geared toward defensive use. The 3-gun crowd has seized this line not only for the same features, but for its reasonable cost. A more basic two-barreled option is available for defensive or sporting roles. Others are strictly short-barreled guns with tactical features like pistol grips, ported chokes, and higher-capacity magazines. A straightforward choice is the "Tactical 5-Shot", with its 18 ½" barrel and breeching choke. The SPX adds a ghost ring sight system and 7-shot magazine. Better yet is the 930 JM-Pro, which stands for Jerry Miculek, a champion shooter. You have to see him in action on a high-speed knock down target course to fully appreciate the combined performance of man and machine. The 930 JM-Pro comes in either 22" or 24" barrel lengths, as a 9 or 10-shot gun. Designed primarily for 3-gun use, controls are oversized. The top-mounted safety is well-located for instant access. Be sure to hit the Mossberg site and watch Jerry's video.

Mossberg's 3-gun Model 930 "Jerry Miculek Pro" offers plenty of fire power.

Winchester. There are still a number of older, gas-operated Model 1400 autoloaders in circulation, most of which employ the "Win-Choke" system. They are a viable used gun choice (if priced right). The new Winchester Repeating Arms "Super X-3" is now in various configurations. The SX-3 employs aluminum alloy to help reduce weight. Synthetic stock versions can be tweaked with shims and recoil pads for individual fit. Like Winchester's latest pumps, the same Invector Plus choke tubes are used to accommodate a large, back-bored 0.742 barrel. Slug shooters will want the "SX3 Cantilever Buck" with its 22" rifled 12 or 20-gauge barrel. A 24", smoothbore turkey model is set up similarly, offering more versatility. The "Coyote" is another interesting derivative with a pistol grip stock. The "Black Shadow" is a more all-around field gun, built to sporting specs. Accessory barrels are available, including rifled types for slugs.

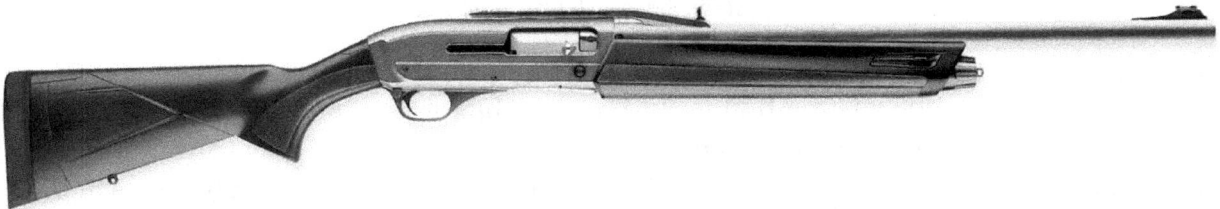

This Winchester SX3 "Cantilever Buck" has a 22" rifled barrel and other versions are available. The SX line has lots of good choices.

Browning. Those facing threats from low-flying aircraft, or possibly Godzilla, should consider a Browning "Gold" 10-gauge. Others would be better served with a lighter and less powerful gun. Fortunately, different models exist. There is a pretty "Silver" series. The "Maxus" is an innovative workhorse, with several nice features. Disassembly is very convenient thanks to a quick take-down lever, eliminating the chore of unscrewing a magazine cap. A clever QD magazine spring retainer makes plug removal fast and simple. As for loading, inserting the first shell into the magazine forces the gun to transfer it into the chamber. Unloading the magazine can be accomplished without cycling shells through the action. A friend shoots a synthetic-stocked 12-gauge, which has seen hard use in harsh conditions (cold weather waterfowl hunts). It feels great and never fails to run. The large safety is appreciated with gloved hands.

Browning's latest offering is the new "A-5." It superficially resembles the original, John Browning designed "Humpback" Auto-5 but is internally quite different. The older gun was the first truly popular semiautomatic shotgun, based on a long recoil action. The barrel traveled rearward from the recoil, driving its bolt. The new A-5 also harnesses recoil, but works like a Benelli. With no gas system, it's "Kinematic Drive" is a very clean system. It also captures many of the gas-operated Maxus features. Between that gun and the new inertial model, Browning can satisfy two main markets. Both 3" and 3 ½" versions are sold to a primarily sporting market. As such, many bolt-on extras aren't available.

Browning's 12 Ga. Maxus gas gun offers handy disassembly and 3 ½" Magnum reach.

The intriguing new Browning A-5 Stalker, which uses an inertial system. Reviews have been positive.

FNH USA. We noted the Browning connection, which began with the first really successful autoloader, the famous recoil-operated, Humpback Auto-5. FN still builds semi-autos, but the interesting SLP Series MK-1 is a gas gun. It's set up for 3-gun, with an extended 8 round magazine and low-profile sights that are cleverly built into a cantilevered scope mount. Invector choke tubes are standard, so it could easily handle most shotgun requirements. Gas operation results in a soft- shooting gun, too. The MK-1 has a synthetic stock with built-in sling studs so it's well set right out of the box. Extra barrels are available in longer lengths.

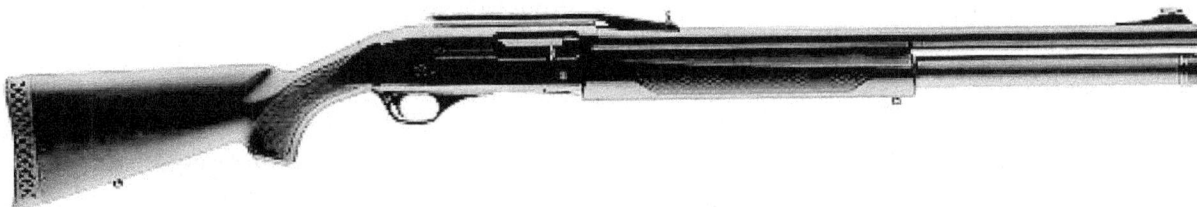

The FNH MK-1 is another interesting cantilevered choice, similar to the Winchester, but with a choke tube equipped smooth-bore barrel. It provides interesting defense or big game possibilities.

A final autoloader choice. The list could easily continue, and many more good guns, but we need to draw the line somewhere, so two final choices follow. Both are well-established designs and available as 12 or 20-gauge guns. We shouldn't forget our basic criteria. Widespread distribution improves the odds of finding parts. Some people counter this concern through a belief that really good guns shouldn't break in the first place. Any mechanical device can fail, and parts can be lost. If cleaning is ever needed on the fly, a bolt handle, spring, or magazine follower can easily disappear in brush. Remembering Murphy's Law, it's best to be prepared.

➢ **Remington Model 1100 or 1187:** As fan of auto-loading shotguns, even though I don't currently own one, a Remington may be your best bet for those pinching pennies. These guns are all over the place and even share some Model 870 parts like safeties, magazine internals, and choke tubes. Spare barrels are common and affordable, as are aftermarket accessories. Recoil is manageable thanks to the gas-operated design. Function is contingent on two piston rings and a large pressure-sealing O-ring, which surround the magazine tube. These parts can be lost, but they're affordable and easy to keep on

Bottom view of a Remington M-1187, showing the rectangular button that must be depressed while loading the magazine.

hand. Cleaning is easy. Loading the magazine is a bit "fiddly", since a button in the bottom of the shell carrier must be depressed while, at the same time, pushing the carrier upward. Only then can a shell be inserted into the magazine tube. I owned an aftermarket device that combined both tasks, but the carrier then jutted out below the receiver. It protruded just far enough to inadvertently drop the bolt upon accidental contact. Periodic replacement of the recoil spring helps prolong life by minimizing battering of the receiver through excessive bolt speed. The factory arrangement really isn't too bad once you get accustomed to it. With regular maintenance and average loads, these guns will have a service life of around 35,000 shells or more. Some competitive shooters will burn through that many shots in one season but, for most folks, that's a lifetime of shooting.

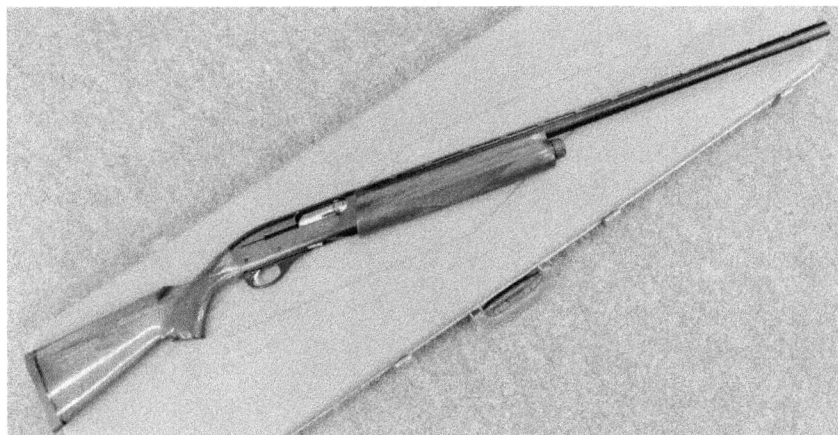

Remington Model 1187 Premier 12-gauge. Less fancy versions are available with dull finish and synthetic stocks.

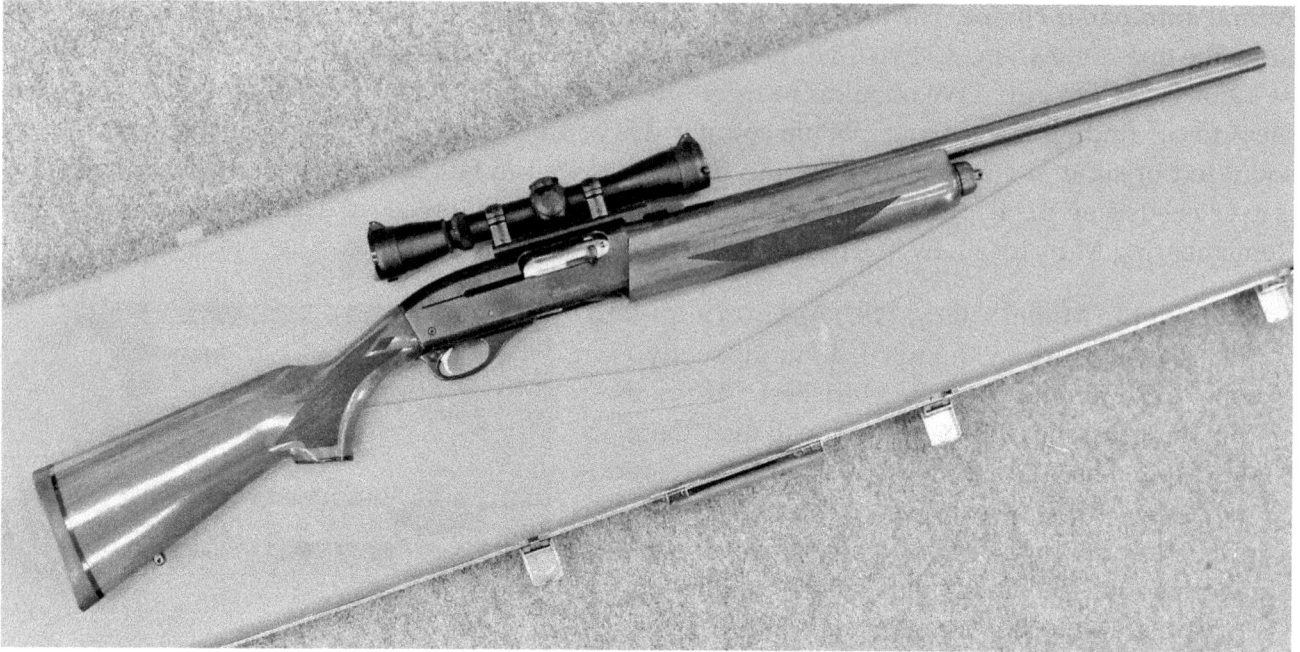

Instant rifle: The cantilevered scope mount ensures constant scope zero when the rifled barrel is attached. This combination will put 3 premium sabot slugs inside 2" at 100 yards!

➤ **Benelli M-2:** Returning to the argument that a good gun shouldn't break, an M-2 will probably be your best bet. It costs more, but you will get what you pay for. Its inertial design runs clean and keeps maintenance simple. Ergonomics are very good and the latest stock design goes a long way toward reducing recoil. Upon disassembly, a Benelli exudes quality throughout. Unlike a Remington, it may not cycle lighter shells, the recommended minimum being a 3 dram load. That said, I recently shot Winchester AA Extra-Light 1-ounce shells through a 26-inch M-2 without a hitch. Recoil was mild and even Winchester's warmer "Super-Handicap" trap loads were

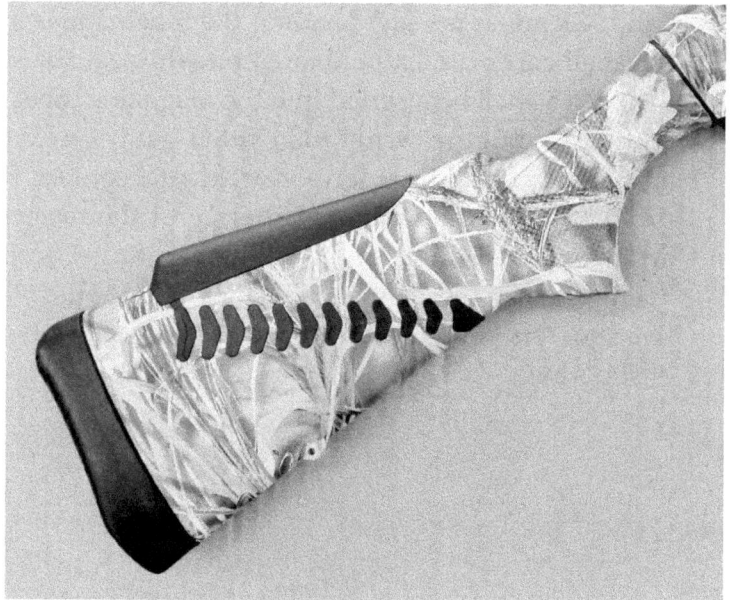

Benelli's "Comfor-tech" stock, which actually works as advertised.

comfortable thanks to the Comfortech stock. One downside to this great gun is that it's somewhat harder to find than a Remington, the latter being common in most gun shops. Parts could be an issue in a time of need. Planning ahead, you could stock up on a few (see the accessories section). I've heard stories of M-2 shotguns that are still going after 500,000 shells. Again, periodic spring replacement helps. The biggest drawback is cost, an M-2 being nearly twice price of a Remington.

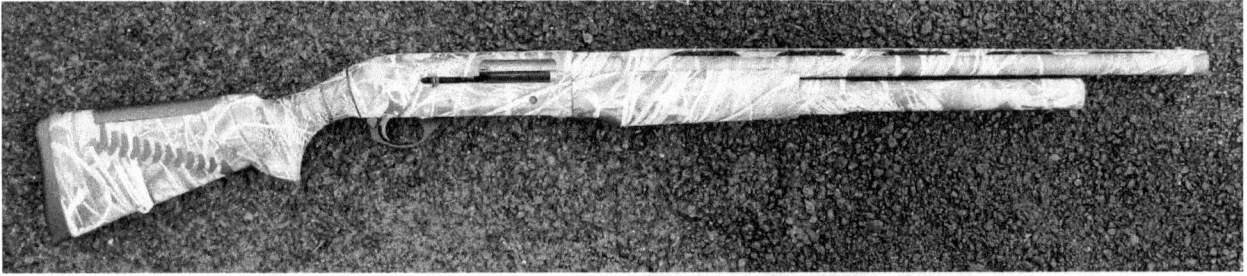

A Benelli M-2 12 Ga. with 26" barrel and magazine tube extension.

Either gun will accept accessories like extended magazines or side-mounted shell holders, as well as spare barrels. Many different Remington and Benelli variations are offered, including models designed to fit small-statured shooters.

OTHER SHOTGUNS

Some very unusual shotguns have hit the market. A few of the strangest are the Kel-Tec KSG, UTAS, and SRM. They use dual tubular magazines and are built for defense. Capacity can be as high as 16 shells! They look formidable and their futuristic appearances have a certain draw. We don't have any experience with a UTAS or SRM, but we do have a small inventory of Kel-Tec KSGs.

The business end of a KSG. Note the twin magazines, tactical choke and RLS light.

Specialty guns. The comments that follow apply to many of these non-conventional, tactically-oriented shotguns. The KSG shotgun is really just a futuristic pump gun which feeds shells from either of its two 6-shot magazine tubes, depending on a selector setting. The magazines are side-by-side and located beneath the gun's short 18 ½" barrel. The KSG employs extensive use of polymers, and the entire package is surprisingly compact due to a Bullpup design that locates the action toward the rear. A pistol grip and fire control system is located ahead of the receiver, and the KSG's action is a modernized Ithaca Model 37 that feeds and ejects from the bottom. Despite its short length, a KSG holds 13 shells! It is not conducive to most sporting purposes. Since most of its weight is concentrated in a fairly small area, it handles like a pig on a shovel. The design also locates the shooter's eye high above the bore and many users employ some sort of optical sight, which can be mounted to a Picatinny rail. Our use is confined to less-lethal projectiles and related specialty loads. The twin magazines permit one to be loaded with bean bag rounds, and the other with loud "flash bang" shells. Operation is more complicated than a conventional pump gun, and it requires more training. We've done plenty of testing but the entire system is fairly new, which raises the previously identified concerns with recent introductions. Our agency's use is limited to highly trained operators. Most people would be much better off with a basic gun for less money. The reason I mention it is to placate any readers who would wonder why I didn't.

The very unusual slide-action Kel-Tec KSG shotgun. This one employs an Aimpoint sight.

Twin magazine followers and selector lever. The bolt is closed on the chamber above them.

More shotgun choices. What about all those other gun makes and conventional models not mentioned? The listings are far from complete. Other brands include the economical H&R Pardner pump, and repeating shotguns of all types from Charles Daly, CZ, Escort, Stoeger, and Weatherby. Most of these guns are made offshore, Turkey being a major new source. A vibrant gun-making industry has sprung up there, using modern manufacturing equipment. Labor costs help keep prices low, while quality is generally good.

I just haven't used them all (or don't know someone who has) long enough to form a well-grounded opinion of how they'll fare in harsh environments. Just about every system has its idiosyncrasies and extensive use is often necessary to validate an example. If you feel confident with another model, great! Still, put it through the wringer. If your life is on the line you don't want any unpleasant surprises; you'll want to know it will work.

PUMP VERSUS SEMI-AUTO

As I've mentioned, there isn't a pump gun in my personal collection. I'm built like a fire hydrant and just don't like the extra reach needed to comfortably run one. Mind you, this doesn't mean I don't know how to. My issued M 870 has gone through a truckload of shells, many of which were fired at a rapid pace. I wouldn't hesitate to take it into harm's way, and have done so on several occasions. But for those of us who shoot for a living, recoil is a constant companion. Some testosterone-laden guys seem to think kick is a necessary test of manhood. We've noticed it's also a great way to develop a nasty flinch. Nobody looks all that manly after a bout of serious missing. The gas-operated autoloaders are my ticket to comfort, explaining their popularity among high-volume target shooters. Additional control is gained by a stationary support hand, which helps when engaging multiple targets.

We've done some speed shooting trials on knock down steel with a remote video camera located downrange to safely face us. The difference in control between a pump and semi-auto is clearly evident when reviewing the footage. Far more gun motion is visible with our pumps. I often hear it's possible to fire one as quickly as an autoloader. For a very select few professional shooters, maybe. We've also seen a lot of hype from manufacturers touting their guns as the fastest. That's fine and dandy, but the object is to make hits. Recovery from recoil takes a bit of time, regardless of the system. Truthfully, in a practical sense, either system will probably do the job, although the self-loading system does promote smoother shooting. The 3-gun shooters run 'em for exactly that reason.

Semi-auto reliability. All is fine and good as long as regular maintenance occurs. Auto-loading bolt assemblies get shoved back into battery by springs and fouling slows things down – a concept not grasped by all shooters. Even if cleaning is regularly performed, a good many shotgunners don't realize that periodic recoil spring replacement is also important. I'm on my third AL-390 spring and can tell when it's getting tired. Slower bolt return and further ejection are always good clues. Ignoring these symptoms will result in stoppages and premature wear.

For those unwilling to deal with such mind boggling issues, simple is better, making a pump the better choice. It will run as clean as an inertial semi-auto, for one third the price. It'll be more reliable in the toughest conditions, and can digest shells of varying power. For any type of societal collapse, I'm grabbing my battle-scarred Model 870, a bunch of 00-buck, and a few slugs.

SUMMARY

Survival Guns: A Beginner's Guide mentions some experimental steel plate competitions between sub-machine guns, pistols, and rifles. I'd be remiss by not mentioning our shotgun experiences. For many years we've made it a practice to drag my old issued M 870 out on the final day of our week-long basic pistol school. We'll have the class elect their two best shooters, who face an array of five knock down targets at twelve yards. A trained pump gun shooter (usually an instructor) faces an identical setup. Starting from a ready position, the object is for the two pistol shooters to defeat the solo shot-gunner. Those not firing serve as judges and the experience is educational. We give it three separate tries, just to be fair. All three shooters are very busy people and the shotgun almost always wins - often by a wide margin. In case you're wondering, we've also put a 9mm submachine gun up against a pump gun with the same target layout. Guess what: The shotgun wins.

One caveat worth driving home involves distance. Results may change as the range increases. We just need to understand the intrinsic limits of each system so we can select the best tool for the job at hand.

CHAPTER 7

ACCESSORIES

There are plenty of interesting shotgun extras available, depending on your gun. We'll examine a few items, along with their plusses and minuses. Just remember that each one adds further expense to the final cost of your firearm system. Weight can quickly become another problem, so the K.I.S.S. factor (Keep It Simple Stupid) has merit. It's easy to go overboard, but a few items may prove useful.

Everything you add increases weight. It's obviously not a problem for this operator, but it can be a real issue for many shooters.

When it comes to defensive use, the lion's share of aftermarket accessories are sold for Remington Model 870 and Mossberg Model 500s, with Benelli M-2 and M-4s close behind. Anyone looking to trick out their shotgun should buy one with this reality in mind. Extended magazines and receiver-mounted shell carriers are two prime examples. Other items like carry cases, choke tube containers, and slings are much more generic.

Slings. With most shoulder-fired guns, you'll only have one free hand unless you either lose custody of your firearm or employ a sling. About the only place a sling may be a liability is in a vehicle, where it could snag on something during fast defensive dismounts. A sling can also be a bother in a safe full of guns, so a QD type is really worthwhile. There are tactical types on the market, and others that accept additional accessories. I can live just fine with a basic nylon type using "Super Swivels." These provide simple pushpin attachment, plus a threaded safety drum that prevents accidental detachment. The sling itself may have

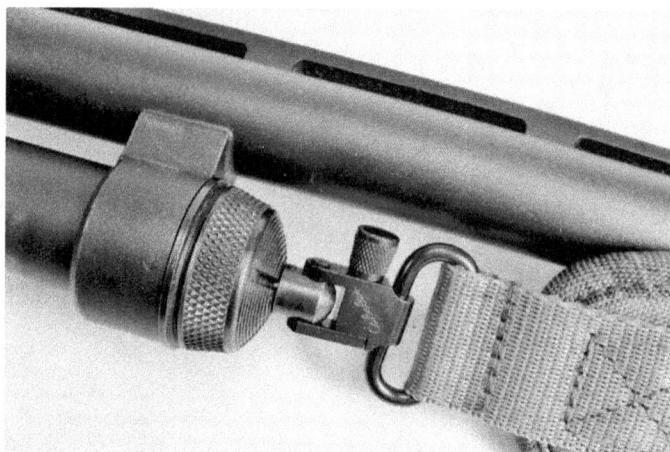

Replacement M-870 magazine cap with QD sling system.

a strip of rubber stitched into the shoulder area and you can buy one with the swivels already attached. Many repeating shotguns are factory-equipped with corresponding mounting studs. If not, the more popular types can be fitted with aftermarket solutions. You can buy a replacement magazine cap for a Model 870 that has a pre-installed stud (some extended magazine tubes may already have one). The rear stud may need to be installed in the stock. It's a fairly basic process, but one best left to a gunsmith for most folks. Cost is fairly reasonable and the same shop can probably set you up with everything. Also, some sling mounting systems use a different QD plugin socket swivel arrangement. It's no accident that a sling is listed first. My double-guns are the only ones without one. They are almost always carried in port arms, ready for a flushing bird. For everything else, I want a sling. Some people hook on shell carriers, which are actually a liability in any dynamic situation. Your shells need a better system.

Depressing the center button on this swivel permits instant hook-up.

Side saddle shell carriers. These mount directly to the receiver and hold spare shells. Once installed, you gain a handy, grab and go, bug-out package. The armory full of M 870s I manage all sport Mesa Tactical 6-shell units, machined from solid aluminum. They're well-made and not prone to cold weather breakage (unlike some plastic carriers). We went to side saddle shell carriers because we couldn't maintain a standard, belt-carried shell bandoleer. Manufacturers would drop or change their products resulting in a mishmash of carriers, all of which occupied essential duty belt space. In the end, and with misgivings, we bolted our spare shell carriers directly to our guns. Our users like 'em, but some receiver-mounted carriers and guns may create issues:

➤ First off, you normally can't just drive the trigger assembly pins out, which complicates maintenance (the pins are replaced by mounting bolts that require tools).

- ➢ Second, the gun gets noticeably heavier.

- ➢ Third, the shells are exposed to the weather which could cause real problems during severe conditions like freezing rain.

- ➢ Fourth, some longer, sporting forends may need modifications to clear the carriers.

- ➢ Fifth, over-tightening of the through bolts may affect gun function. We've heard of recoil-related issues resulting from soft aluminum receivers, but our steel Remington guns are holding up fine.

On a positive note, these devices provide a good ammo-management system for organization of different shells like slugs and buckshot. The biggest concern seems to be whether to load them with shell bases up, or down; to improve loss prevention, we load them up. These units have merit for a dedicated fighting gun; however, I'd skip 'em on a general utility shotgun in favor of a compact belt carrier. Your gun will handle much better and be easier to service. Besides Mesa, Advanced Technology, GG&G, Tac Star, and Pacmayer all sell these devices. Some cleverly incorporate a Picatinny rail for mounting optics. Read their specifications carefully to ensure a proper fit.

Mesa side-saddle shell carrier in action.

Another version with a slug at the rear. Note the sling mounting system attached to the Picatinny rail on this Mossberg shotgun.

Other shell carry methods. The most popular spare shell device may be an elastic carrier, designed to slip over a buttstock. I have a truck carbine so equipped, but it lives in a case primarily for varmint control purposes. Long-term exposure to weather isn't an issue, nor is the possibility of opposite-side shooting. They're inexpensive and you gain a grab and go package. The loops eventually get tired, so they need periodic inspection and replacement. Ice and snow cause other problems, but even without them, shells can corrode from accumulated moisture. Most types also slide onto a belt, where they can be shielded by a coat. If located on the forward support side, effective reloads are possible while maintain a firing grasp of the gun.

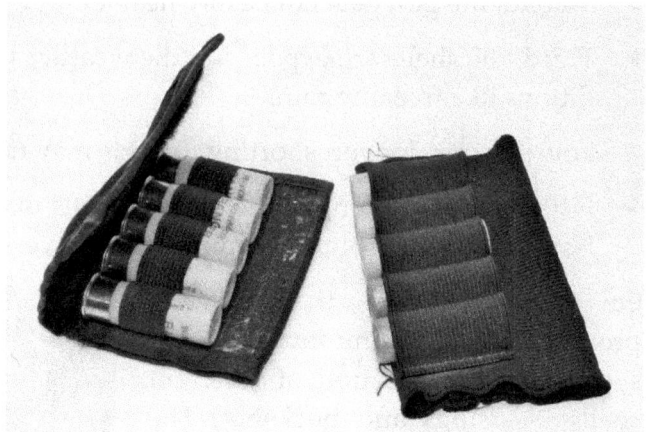

Elastic belt and stock-mounted shell holders. Note the pair of different shells. They're slugs.

Safariland sells a neat little plastic two-shell belt holder that does the same thing. A modified version is designed to screw into a stock and provide fast access with reasonable retention. These devices also provide a means to segregate different shells like buckshot and slugs. We've also used belt carriers with Velcro flaps that offer protection from the elements. Most of these items are inexpensive. The 3-gun crowd has spawned development of all sorts of fascinating gadgets from tubular speedloaders to open-top, belt-mounted, fast-access carriers. A few innovative products will mount to a gun for instant access of spare shells. In the real world though, think "loss" or "weather." The latter is a bigger deal in northern climates, where two feet of snow can invoke both problems. If ammo is scarce, every extra shell will be precious!

We've played with large-capacity bandoleers. You get plenty of shells, but the gun also gains weight. At least it's distributed around your body, which can't be said of trouser pockets. A belt-carried sidearm won't work well with a full bandoleer and you'll look like Pancho Villa. I'd much prefer a sportsman's game vest with heavy-duty pockets if lugging a full box of shells. One interesting modification we saw involved a flexible 5-shell carrier with a Velcro

Safariland dual-shell holders with a three-gun competition device.

back. It permitted QD receiver attachment, using a second adhesive Velcro backer mounted to a shotgun. The user could store it on his/her shotgun, or pull it off and slap it on a tactical vest. Besides providing a way to carry extra shells, a well-thought-out carrier permits sequential access to tactically support the system.

Extended magazines. Remington's standard-length, 4-shot magazine tubes often have opposing indents near the cap end to capture the spring during barrel removal. The two indents also prevent passage of extra shells if a magazine extension is installed. Some folks drill them out and I've swaged a few back to tube diameter. Since adding an extension kills the QD barrel feature, I'd skip it unless use was confined to defense applications. In that case, I'd consider one of the dedicated defensive guns with rifle sights, synthetic stock, and an interchangeable choke system. An extended magazine adds considerable weight. Between one and a side saddle shell carrier, a law enforcement type Model 870s is tough to hold in a high ready position for more than just a few minutes. As equipped for my agency, these accessories support 13 shells; lots of firepower, but plenty of heft! If you stick with the standard 4-shot magazine all is not lost. A small belt-carried bandoleer located on your non-gun side will permit sustained loading. You can shoot and replace with your support hand while your firing hand controls the gun. With practice, it's highly effective and overcomes the limited magazine capacity issue. A good source for extended magazines is Briley Manufacturing. They also sell tube-mounted mounting points for lights and slings.

Rem M-870 shown in "high ready" with 7-shot extended magazine.

Aftermarket magazine followers. The factory unit is pretty cheesy. We switch them out for lime-green colored Scattergun Technologies; solid, plastic units. They're rugged, and a little nub molded into the follower's face provides tactile feed-back in the dark. If you can feel this protrusion, your magazine tube is empty. We heard a story about an agency that experienced a shell detonation after a shooter loaded a shell backwards into a magazine. Shells were inserted until the tube was full. When the gun fired, inertia caused the nub to act like a firing pin, discharging the shell. This could be just another urban myth, but we did file small flats on our followers after hearing about it. All in all, they're a trouble-free and worthwhile accessory considering their negligible cost. TacStar sells one, and GG&G also markets a nice-looking stainless steel unit, which can't imbed debris. It has corresponding indents that permit insertion in the dimpled Remington tubes.

Scattergun Technologies high-visibility magazine follower with tactile nub.

Ghost ring peep sights. A ghost ring sight improves slug accuracy which, combined with a rifled choke tube, may tighten your groups. I'm no stranger to the concept and have a semi-auto .308 set up for fast shooting in bad weather. A ghost ring is really just a peep sight with a large aperture. The big opening lets in plenty of light while permitting fast sight alignment; so on a fighting shotgun, a large peep sight makes lots of sense; but a few tactical operators I hold in high regard have found a big peep to be a hindrance during close-quarter engagements. With some regret, we've come to the same conclusion: It boils down to your intended use and personal preference.

A serious fighting gun, Benelli's M-2, equipped with ghost ring sights.

XS Shotgun Express Sights. Here's a possible alternative to a ghost ring, providing a much smaller and uncluttered profile. An unobtrusive rear sight goes in a ventilated rib and a radioactive, glow-in-the-dark Tritium bead is installed up front. This is a low-profile arrangement, which better lends itself to wing shooting. It could offer improved accuracy over a standard arrangement with slugs as well. I once set up a custom M-1100 vent rib barrel with a similar arrangement and it worked fine. The XS system requires a gunsmith because the rib will need machining to mount the rear blade. A 26-inch bird barrel might be the right length – not as short as most slug barrels, but long enough for flying targets. You might try a rifled choke tube to see if accuracy improves. Reports vary, with slower-velocity slugs appearing to be the better choice.

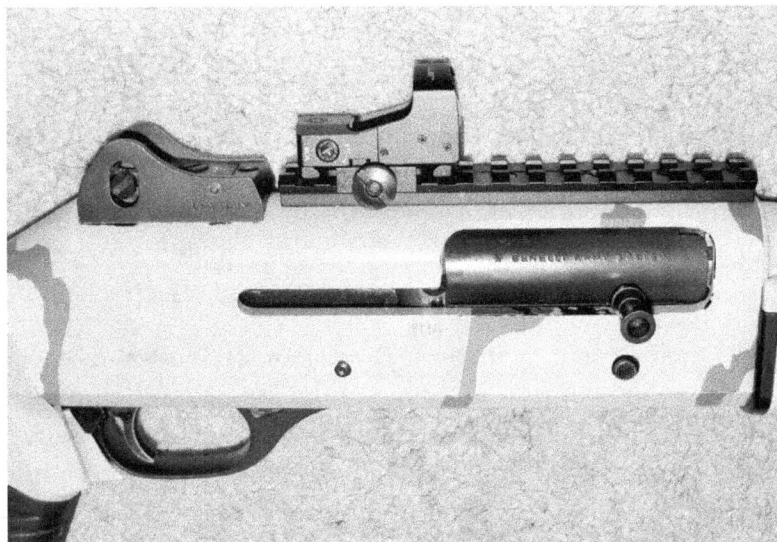

A compact Burris Fast-Fire III mounted ahead of a ghost-ring sight on the Picatinny Rail of a Benelli M-4. The optical sight can be detached with a quarter if need be.

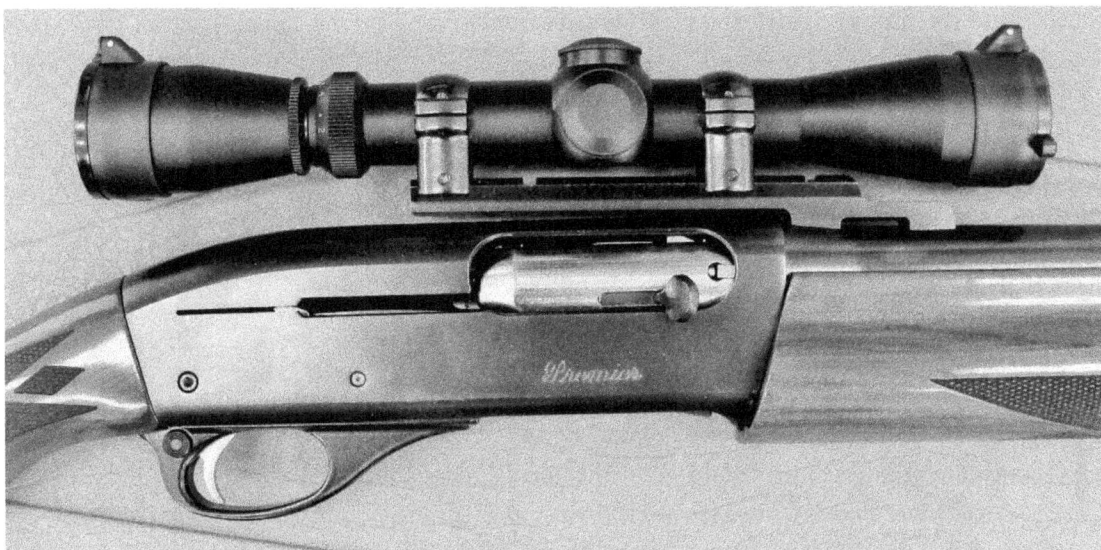

Serious deer medicine: Cantilever scope base welded to a Remington M-1187 rifled barrel topped with a Leupold 2x7 Shotgun Scope. Zero remains constant.

Optical sights. Variations include non-magnifying dot sights or scopes. The comments about ghost ring peeps generally apply to optics, with a few exceptions. Many turkey hunters prefer the precision of optical sights. For extended-range use on big game, a fully rifled slug barrel can be purchased that has a cantilevered scope mount. The mount is welded to the barrel and extends rearward, above the receiver. You gain repeatable zero with this arrangement because your optic remains attached to the barrel. In essence, you gain a switch-barrel shotgun and scoped "rifle" set. This seems preferable to bracket-type saddle arrangements that semi-permanently attach to the receiver. Those, and the hybrid shell-carrier/Picatinny mounts, will need to be detached at some point. Some loss of zero is possible and this prospect tends to discourage trigger group servicing. Another drawback is the height of your aiming device in relation to the barrel. You'll probably need more stock height to permit a proper gun mount. Several products will provide a simple remedy, but not completely quick-detachable.

While hosting a military range program, an assortment of Remington M 870s appeared that had Trijicon Reflex dot sights. We're very familiar with these rugged, battery-free units, which we mounted on an armory full of AR-15s. Unfortunately, in the case of the shotguns, overall handling qualities were lost due to a poor aftermarket stock choice with too much drop. The result was a very awkward package. The sum of all add-on gadgets will dictate handling qualities for better – or in this case, for worse. In smoothbore land, K.I.S.S. supports the goal of fast and effective shooting.

Giant head safeties. I like these. You may increase the risk of inadvertently knocking one to the "fire" mode, but the jumbo button is easier to find. Some of the later Remington safety buttons have key lock feature, which is lost when switching to the aftermarket model. It's really your call concerning the level of security necessary. The "J-Lock" contains a small locking mechanism within the actual safety button. Remington only offered it for a few years and it's now discontinued. Missing keys may have been part of the reason, although YouTube has videos that show other ways to unlock them. I've had good luck with a basic Scattergun Technologies/Wilson Combat replacement. With a bit of patience, you can perform the installation yourself after driving out the two trigger-assembly retention pins to access its innards. Either version is reversible for use by lefties.

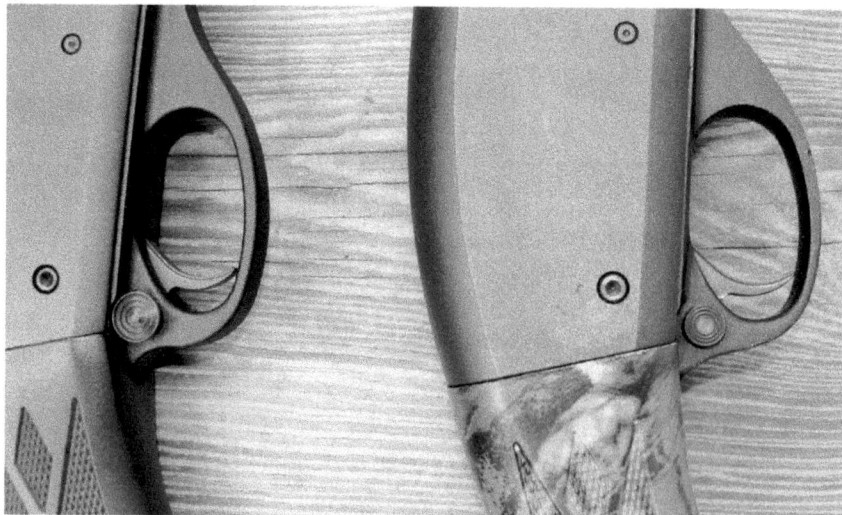

Remington safety and replacement giant-head version (L).

LE Forends. Many of the M 870s you buy over-the-counter will have a sporting forend. It's longer than a law enforcement (LE) version, extending further rearward to accommodate users with a shorter reach; however, when the action is fully opened, the sporting forend covers much of the receiver's lower magazine access cutout. Unloading the magazine is typically done by shucking shells through the action. There is a better way! Our LE M 870s have shorter forends that don't obstruct magazine access when fully rearward. Our troops are trained to unload their magazine tubes using the shell stops. Depressing the retaining tabs will allow shells to pop out of the magazine. This procedure is safer, because shells never travel through the action. Instead, once the correct technique is mastered, they just pop out the bottom opening. It's pretty slick. I've shortened a few sporting forends using a band saw and sanding disk. It's a simple modification for either wood or synthetic versions. One thing to watch out for though is more difficult operation with short-statured shooters. The shorter version requires more reach. Last thought here: You can buy an LE-type forend with an integral light – something to consider if your pockets are deep. Surefire has a beauty. Forends like the Mako Group unit, with short rail sections attached, are available for much less money. Changing a forend on a pump will require some sort of tool (see the maintenance chapter).

A shorter LE forend permits safe unloading with the action open. Just manually trip the shell stops and empty the magazine while the slide stays fully rearward.

A tricked-out Mako forend with Streamlight and extra Picatinny rail sections.

Lights and mounts. At best, you'll be a one-handed shooter when holding a shotgun. Trying to manipulate a hand-held light is difficult, even without pumping a shotgun. A forend with an integral light is a nice addition. In essence, you'll be paying for a new forend and a dedicated light, whereas rail-mounted versions may permit mounting of lights already on hand. I like my Safariland RLS lights, which quickly mount to any Picatinny Rail by just the twist of a hinged base. The latest version runs on three AAA batteries but throws 90 Lumens (it's bright). The Mil-Spec Picatinny system has become extremely popular on AR-15 forends, but accessory rails are gaining ground. Anyone with an extended magazine should have enough room to accommodate a clamp-on rail section. The light may then be switched to other firearms, which saves money. I keep one on an AR-15 forend, but a second RLS jumps between a pistol and other rail-equipped shoulder-fired guns. TacStar sells some interesting magazine-mounted lights.

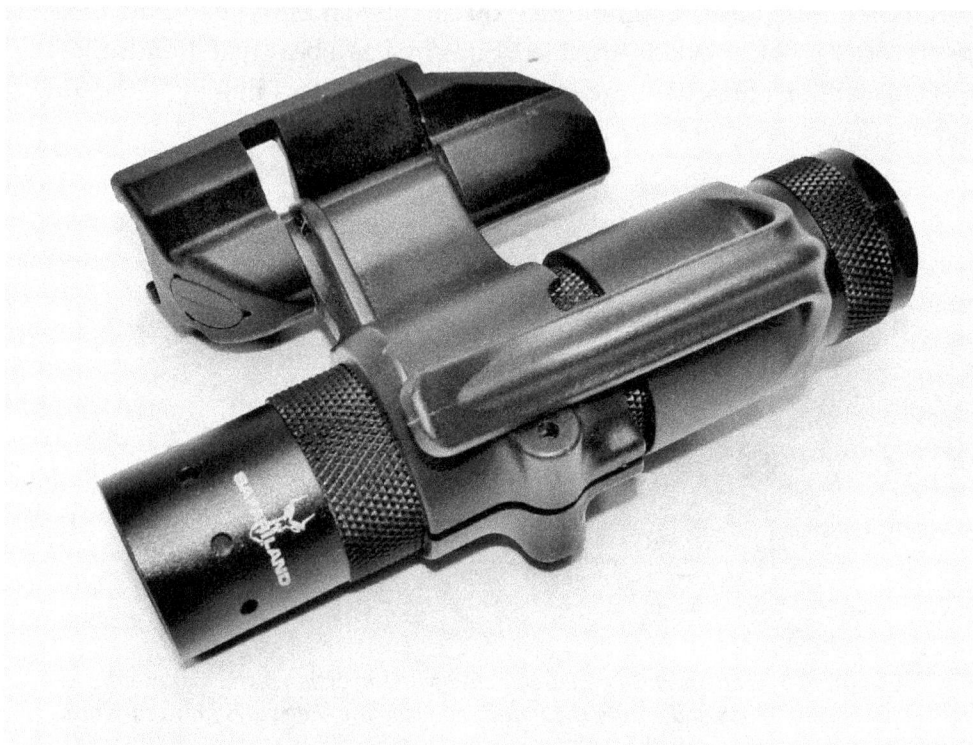

Safariland RLS, which stands for "rapid light system." The pivoting mount permits QD mounting on Picatinny-type rails.

Pistol grips instead of stocks. There's no denying a role for these in very tight places. I once trained with a special operations unit that developed some highly unusual techniques using short-barreled, twin-grip pump guns inside vehicles. In case you're wondering, it was a very loud experience! Another time I watched one of our tactical operators hammer screaming mini-claybirds that were launched from a tower. It was an impressive display of gun-handling beyond the ability of most shooters. Sticking with our general-use theme, I'd opt for a more conventional stock configuration. In low light, or when things go to hell quickly, the opportunity to gain peripheral barrel reference may not exist. On the other hand, with reasonable gun fit and a solid cheek weld, you can expect patterns to impact close to where you're looking.

Decked out Benelli with extended magazine, ghost-ring sight, pistol grip, and tactical stock. This version permits a good cheek-weld, but others don't.

This Knox system has a spring-loaded recoil reducer. An accessory cheek-piece riser would help decrease drop.

Ninja stocks. The latest fad involves bolting M-16-type stocks on every sort of shoulder-fired firearm. Several versions are offered for pump guns and some are okay while others are not. Watch out for increased stock-drop, resulting in poor cheek weld. You'll shoot poorly unless your head is anchored to a consistent stock location. One aftermarket stock we like is the Davis Speedfeed. It's more conventional, but locates two extra shells on each side in compartments near the rear, providing positive ammo segregation. With a fighting load of 00 Buck up front, the extra stock-mounted spots offer ready slug access. Reality check: unless you drill extensively on ammo change outs, none of this will really matter when things go south.

Extra barrels. Again, if I could only have one gun, it would be a Remington M 870 12-gauge chambered for 3-inch magnum shells. I'd want at least two barrels: a 26 (or 28)-inch bird barrel, and a 21-inch smoothbore, rifle-sighted deer barrel. Ideally, both would be interchangeable choke tube versions. Either could be switched in seconds without tools to provide shotgun or rifle-like performance. Choke tubes would expand the usefulness of the rifle-sighted barrel. A modified tube would throw great 00 Buck patterns and a full choke would be the ticket for turkeys or predators. For true rifle-like performance, consider a third, fully rifled slug barrel with a cantilever scope base. These are designed for use with optical sights that mount directly to the barrel. The rifling permits use of sabot rounds, which expel plastic-sleeved, sub-gauge projectiles at high velocities. Winchester catalogs a 300 grain XP3 slug with a velocity of 2100 fps so, in essence, you'd have a fast-handling .45/70 pump gun. Whatever optical sight is attached to the cantilever base will remain zeroed. Viola, instant rifle! Standard Foster slugs will lead them up but sabot-type projectiles can deliver impressive results well beyond 100 yards. The drawback: Shot patterns are blown badly. That's why you still need at least one smoothbore barrel. With all three, the wide assortment of munitions commonly available would provide a very flexible system.

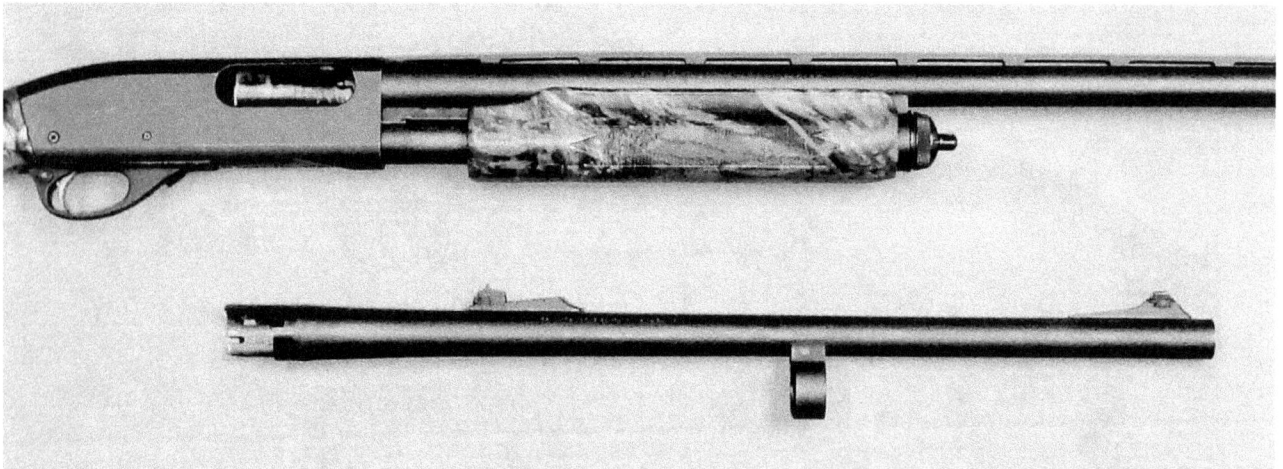

Extra M-870 21-inch slug barrel with rifle sights. This one accepts Rem-Chokes.

Silenced shotguns (the Metro barrel). Among my Mobilchoke assortment sits the strangest tube one could imagine. I first got a close look at this strange device in a friend's kitchen. An unusually long Remington M 870 was leaning in a corner and I almost poked it through the ceiling when I picked it up. Picture a 34-inch choke tube threaded onto a standard barrel and you'll have the right idea. Although not the ideal choice for room clearing, the extension can be very quiet with the right loads. A series of radial ports bleed off pressure and the Metro barrel is threaded at both ends. After unscrewing the choke from your standard barrel, the Metro barrel is hand-turned until snug. The muzzle end is threaded for the same factory choke tube just removed. The Metro barrel doesn't contain baffles or any exterior housing, and isn't classified by BATF as a silencer. Still, it would be worth checking your state's laws before bringing it afield.

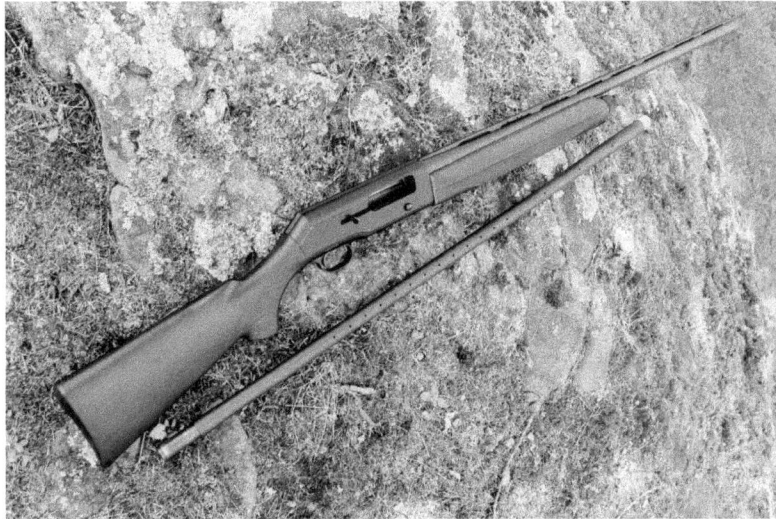

Beretta AL390 12 Ga. auto with super-quiet Metro Barrel extension.

I managed to track down a used Beretta Mobilchoke version and dragged out a few boxes of Winchester "Feather", low recoil/low noise, #8, 12-gauge shells (which act more like watered-down 20-gauges). This load is one of a few recommended for the Metro barrel, another being Federal's Metro subsonic #7 ½, specifically designed for this device. The improved cylinder Mobilchoke normally installed in the muzzle of my Beretta AL 390 was switched out for the Metro barrel and the I/C tube was relocated to the Metro's business end. I also replaced the standard AL390 gas valve spring with a light load version. Function with the "Feathers" was 100% and the fun began.

Low recoil / low noise 12 Ga. shells, which perform as advertised.

Handling was awkward to say the least. I practiced swinging the gun outdoors to prevent breaking light fixtures and scratching ceilings. Woodcock hunting was clearly out, but in open spaces the gun was manageable. Hearing protection wasn't necessary and from a few feet away, reports sounded more like a big-bore airgun. The system took some getting used to, but was surprisingly effective once mastered. We had lots of fun smashing claybirds and I used it on a crow control hunt with great success. With our normal loads, the wary birds vamoose as soon as the shooting begins. With the Metro barrel, they continued to circle our decoys long enough to sustain several volleys. Flights arrived more frequently, probably because the surrounding area wasn't shaken up. Range was shortened due to the reduced loads, but the system was still effective out to 35 yards using an IC choke.

Metro Barrel installed: Quiet, yes, but not the ideal brush gun system.

I had to worry about reliable semi-automatic function, but this concern is eliminated with a pump. Assuming you have a choke tube-equipped 12-gauge Remington Model 870, you can just thread on the correct model Metro barrel and go to work. The system has value for anyone concerned about tell-tale reports. Federal and state animal control agents use them in some surprisingly populated areas like airports, golf courses, and parks. Geese are common targets, but subsonic slugs can also be fired for use against large animals.

The now defunct manufacturer of aftermarket shotgun barrels, Hastings, produced Metro barrels and mine bears that label. When Hastings closed its doors the supply dried up, but a new Metro Gun website is up and running again. It's worth a look if for no other reason than curiosity.

Spare choke-tubes and wrench, which fit nicely in the simple MTM case.

Other extras. You'll want a few choke tubes if your gun is so equipped, plus a small plastic case for storage. Prices for an actual gun case can vary greatly. We use floating, waterproof soft cases in our duck boats for obvious reasons. A takedown case is handy if compact storage is desired. You'll also need cleaning gear, which we'll cover separately. A few dummy shells are handy for training purposes, and we'll expand upon them later on.

A serviceable collection of shotgun accessories.

Sources. Brownell's is a great resource for accessories and gunsmithing supplies. They've been in business 75 years and built their company on good old-fashioned customer service. Midway USA is a bit younger, but well-established. They offer tech support and can help you out with questions pertaining to fit. Graf & Sons is a good ammo and reloading supply source, but sells other accessories as well. The big sporting retailers like Cabelas, Gander Mountain, and Bass Pro Shops are good for the more generic items. Since they have stores throughout the U.S., hands-on comparisons are possible.

SUMMARY

It's easy to go overboard on widgets and shiny toys, but a few key items should be factored into your total system cost. Fortunately, they aren't too expensive. Among them are a sling, shell carrier of some sort, carry case, and cleaning gear. The other stuff can be procured piecemeal later on.

CHAPTER 8

USEFUL SHOTGUN LOADS

The great attribute of a shotgun is its ability to fire a wide array of ammunition including birdshot, buckshot, slugs, and other specialty loads from less-lethal through breeching rounds. For general purposes, a few common shells will suffice.

Small game, wing shooting, and training. Birdshot is what you'll need and your quarry will dictate the loads. You can use charts to help narrow down choices and some shell manufacturers print one right on the box. Consider pattern density, governed by range and choke, balanced against the terminal energy of individual pellets. We also use birdshot during defensive shotgun training. Steel reactive targets topple nicely with the right loads for considerably less cost than buckshot. Make sure you're shooting lead pellets. Steel shot will make for a nasty experience!

SELECTING THE RIGHT SHOTSHELL	GAUGE	DISTANCE	CHOKE	LEAD SHOT SIZES
Turkey	10, 12, 20	20-30 YD	F	4, 5, 6
	10, 12	30+	F/EF	4, 5, 6
Pheasant, Prairie Grouse	12, 16, 20, 28	20-30	IC/M	4, 5, 6, 7½
	12, 16, 20	30+	M/F	4, 5, 6
Ruffed Grouse, Partridge	12, 16, 20, 28	20-30	SK/IC/M	6, 7½, 8, 9
	12, 16, 20	30+	IC/M	5, 6, 7½
Quail Dove	12, 16, 20, 28	20-30	SK/IC/M	7½, 8, 9
	12, 16, 20	30+	IC/M	7½, 8
Woodcock, Rail Snipe	12, 16, 20, 28	20-30	SK/IC/M	7½, 8, 9
	12, 16, 20	30+	IC/M	7½, 8
Rabbit, Squirrel	12, 16, 20, 28, .410	20-30	IC/M	4, 5, 6, 7½
	12, 16, 20	30+	IC/M/F	4, 5, 6

Federal recommends patterning your gun to determine the optimum choke.
EF = Extra Full **F** = Full **M** = Modified **IC** = Improved Cylinder **SK** = Skeet

A handy reference chart containing lots of useful information.

The loads listed below are all 12-gauge 2 ¾" choices and there are plenty of other great picks. We buy Winchester shells in volume and, during the past 25 years, we've fired tons (literally!) of AAs without a single problem. They cost a bit more, but they provide an extra benefit. Because the AA hull is popular with reloaders, good components and data are readily available. Depending on hull condition and load intensity, at least several reloads should be possible from one fired shell. Even though the AA is a low-brass target hull, heavier loads can be developed for it. Some of the same powders work in handgun reloads, which simplifies propellant inventory. We've also had good luck with Federal and Remington target loads. For reloading purposes, the lowest-priced promotional shells don't hold up.

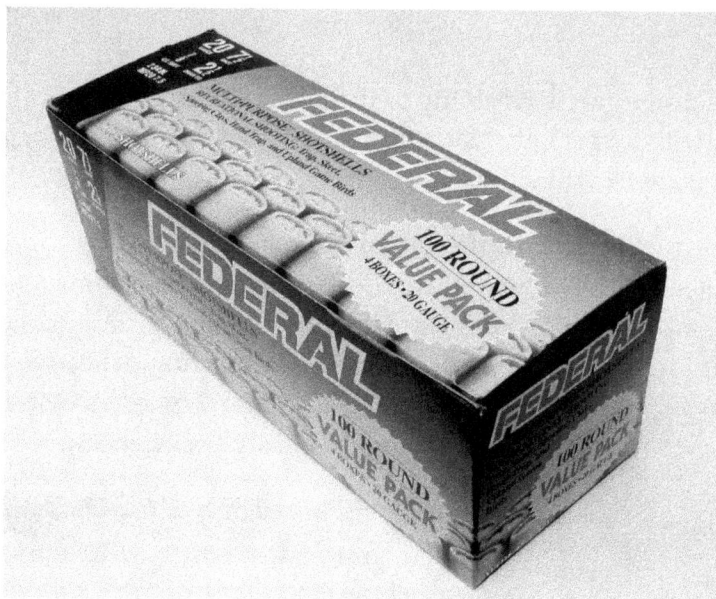

A typical package of so-called "promo shells", adequate for basic shooting.

➤ **Low-cost promo shells:** The lower-cost promo shells sold by big-box retailers will work just fine. They are often sold with #7 ½ pellets. Just be aware that the lightest loads may not have enough power to cycle all semi-auto shotguns. In that case, a 3 dram load will probably be your minimum. Low cost will be the deciding factor, rather than any specific brand.

➤ **Low recoil/low noise loads:** These are light specialty loads. They're a nice way to get a novice up and running. In my Metro Barrel, report is no louder than a .22 Long Rifle. Pump gun shooters can run the lightest loads, including Winchester's AA "Feather" low recoil/low noise shells. This very light 12-gauge load is advertised as having 50% less recoil and noise. The recoil is much like a 20-gauge #8 with an equivalent payload, making it a good choice for smaller-statured shooters. Federal's Metro 1 dram low recoil subsonic load is another nice choice, and uses a heavier 1 ounce shot charge. My gas-operated Beretta M-390 self-loader will feed either, but they lack the power needed to run many other models.

➤ **1-ounce shells:** These somewhat heavier, conventional 12-gauge loads will break claybirds at typical ranges without beating you to death. They also work on smaller upland birds like doves, quail, or woodcock. Smaller #8 shot helps maintain decent pattern density. I have a lot of experience with Winchester AA "Xtra-Lite" 2¾-dram shells. Federal's similar "Target" load works equally well. These shells are a step up from the promo-loads, and cost a bit more, but their harder pellets and better wads also throw more uniform patterns. Save your fired hulls for future reloading, whether for you or someone else.

➤ **1 ⅛ ounce target loads:** The 2¾ dram 1 ⅛ ounce target shells will cover many claybird games. A Winchester AA 2¾-dram "Lite" #8 load is a heavier version of the "Xtra-Lite", and makes a good all-around choice. Skeet shooters often shoot them (or #9s) to gain pattern density on fairly close range targets, using open chokes. The warmer 3 dram # 7½ shells are popular for longer-range targets shot with tighter chokes on regulation trap courses. We use lots of Winchester AA "Super-Handicap" 1250 fps shells because they have recoil similar to low recoil buckshot. The 1 ⅛ ounce target shells also perform well on upland birds as large as ruffed grouse.

➤ **Upland game and small game loads:** Although I've shot plenty of ringneck pheasants with warm Winchester "Super-Handicap" 1 ⅛ ounce #7 ½ trap loads, these birds are tough enough to warrant a switch to larger pellets. A good all-around upland shot size is #6. The best copper-plated high-velocity shells work surprisingly well. Federal's "Wing-Shok High Velocity" lead load is listed at a very fast 1500 fps and the copper-plated #6 pellets pattern very well. Besides pheasants, I've used them extensively during crow control shoots, toppling birds out to 50 yards. In the Midwest, where spaces are open and tough ringnecks often flush wild, many hunters move up to heavier 1 ¼-ounce high brass game loads. Squirrel hunters often use similar shells. Rabbit hunters do too, sometimes shooting larger pellets like #5s or #4s. The latter size is popular in northern New England on snowshoe hares, which are larger than cottontail rabbits.

Some useful Winchester AA loads. Save the fired shells for future reloading.

More good loads from Federal. Note the wide range of listed velocities.

Larger species. These normally require larger pellets for adequate terminal energy and penetration. Pattern density can be maintained by increasing shot payload, and the harder-kicking but longer 3" Magnum shells are a popular solution. One exception is turkeys, usually hunted by aiming for the head and neck. Magnum quantities of smaller pellets are often used to increase hits on the relatively small target area.

Useful non-toxic waterfowl loads in various pellet sizes and composition.

HEVI•METAL™ GAME GUIDE

GAUGE	LENGTH	SHOT SIZE	LOAD	FPS	SPECIES
12	3"	BBB*	1-1/4	1500	GOOSE
12	3"	BB	1-1/4	1500	GOOSE
12	3"	2	1-1/4	1500	DUCK
12	3"	3	1-1/4	1500	DUCK
12	3"	4	1-1/4	1500	DUCK
12	3"	6	1-1/4	1500	TEAL
12	3.5"	BBB*	1-1/2	1500	GOOSE
12	3.5"	BB	1-1/2	1500	GOOSE
12	3.5"	2	1-1/2	1500	DUCK
12	3.5"	3	1-1/2	1500	DUCK
12	3.5"	4	1-1/2	1500	DUCK
20	3"	2	1	1350	DUCK
20	3"	3	1	1350	DUCK
20	3"	4	1	1350	DUCK

HEVI•METAL™ CHOKE CONSTRICTION

SKEET	60%
IMPROVED CYLINDER	65%
MODIFIED	70%
FULL	80%

Pattern Performance VS. Choke Constriction for 12 Ga Guns
(Choke is measured by the percent of
pellets in a 30" Circle @ 40 Yards)

A handy non-toxic reference chart printed right on this Hevi-Metal box.

> **Waterfowl:** Back when lead shot was legal, I used plenty of 12-gauge, 2 ¾", 1 ¼-ounce #6 or #5 shells in a modified choke, killing ducks out to 40 yards. Non-toxic shot requirements changed everything. The first alternatives were steel pellets, which were lighter in any given shot size. After lots of initial hunter confusion, a simple remedy involved switching to 3" Magnums, while moving up two pellet sizes to improve killing power. Like many other hunters, I now use 3" Magnum steel #3s for general purpose, shorter-range duck hunting. The harder pellets require less choke so I use I/C, shortening up the distance somewhat in deference to rapid velocity loss and decreased pellet energy. The latest non-toxic alternatives offer greatly improved performance through heavier tungsten/iron pellet alloys. I shoot these when longer ranges are expected, but their improved performance comes with a much higher price. Some dyed-in-the-wool waterfowl hunters shoot 3 ½" Super Magnum shells to gain maximum performance on tough birds like geese. I switch to a premium 3" Magnum #2 shell, which offers a fair balance of recoil, pattern density, and pellet energy. I've also used larger BBs, which are popular in heavier 3 ½" payloads.

➤ **Turkeys:** Lead shot is legal in most places and pellets are often restricted to a few intermediate sizes, the object being to aim for the head. Ranges can be long, so pattern density is important, as is adequate energy. I've had good luck shooting premium12-gauge, 3" Magnum #5s in a full choke, or specialized extra-tight "turkey" constrictions. This is one circumstance where a shotgun is actually aimed rather than pointed. I've killed plenty of birds using a 28" front bead only bird barrel, but I'm always careful to maintain a consistent gun mount and a properly fitted gun. I've switched to a shorter-barreled gun with a second bead that can serve as a rough rear sight. Died-in-the-wool turkey hunters often use more sophisticated aiming systems, which may even include a scope. While a nice dense pattern is essential, it also needs to center on the spot you're trying to hit! It's worth capturing a few patterns on a target to see what you're working with.

➤ **Larger fur bearers:** The list may include raccoons, foxes, bobcats, and coyotes. Bigger pellets will be necessary since these animals are not only larger, but more tenacious. Lead #2 shot is a starting point for the smaller species, but a large coyote is one tough customer. In our area #4 Buckshot is popular during hound hunts, as is the specialized "Dead Coyote" load.

Effective general-purpose turkey loads in clearly marked boxes.

This turkey was bagged with a bead-sighted gun and standard full choke.

While most of these recommendations apply to 12-gauge shotguns, 20-gauge versions can be substituted. Although payloads will decrease, a 3" Magnum 20-gauge still has reasonable power.

Interior home defense. Consider a dual-purpose shell from the field category above. A 2 ¾" 12-gauge load of #6 shot is devastating at close range, and yet far less likely to blow through walls. Fire a pattern at 5-7 yards for a graphic illustration of what you can expect. Moving up in punch, #4 Buckshot is worth a look – especially if more range is needed.

General defense. You can't go wrong with 00 Buck. The standard 12-gauge, 2 ¾" load contains 9 pellets, each measuring .33 caliber. This load has been the buckshot benchmark for years. We have considerable experience with several brands, including Winchester and Federal's so-called "tactical", low recoil loads. Velocity is a bit slower, but patterns are very good, even from more open chokes. Recoil is on par with Winchester "Super-Handicap" #7 ½ Trap loads, which we use for training on steel knock down targets. As mentioned previously, three-inch loads are pretty rugged in the recoil department. They do contain more pellets (15), but unless you're a tough individual, it would make more sense to stick with the most manageable loads available. I wish I had a dollar for every flinched buckshot discharge I've witnessed, even using standard, 2 ¾" shells.

Although 00 Buck is the most popular choice, the other variants are worth a look. Some people like #4 Buckshot. A 2 ¾", 12-gauge shell contains 27 smaller pellets, each measuring about .24 caliber. They improve pattern density, but quickly shed terminal energy. It probably boils down to your own, unique situation. Shoot some patterns to get a handle on performance from your barrel and choke combination. Be careful not to judge results based on just a few shots. Buckshot can do weird stuff at times, so the more you shoot, the more reliable your conclusions. We prefer Winchester buckshot shells with sealed crimps for regular carry in a vehicle or ATV (check for legality). Granulated plastic buffering compound helps cushion the pellets during discharge, but vehicle vibration can cause it to leak through unsealed crimps. It's not common but it happens, and in a magazine tube, this can cause problems. The stuff looks like coarse white powder, and Winchester melts the plastic crimp petals together so it won't leak out. You can't reload these shells because the small center portion will blow off upon discharge. Then again, most people don't reload buckshot hulls.

One ragged hole: 12 Gauge, 1 ⅛-oz load of #6s fired from 5 yards.

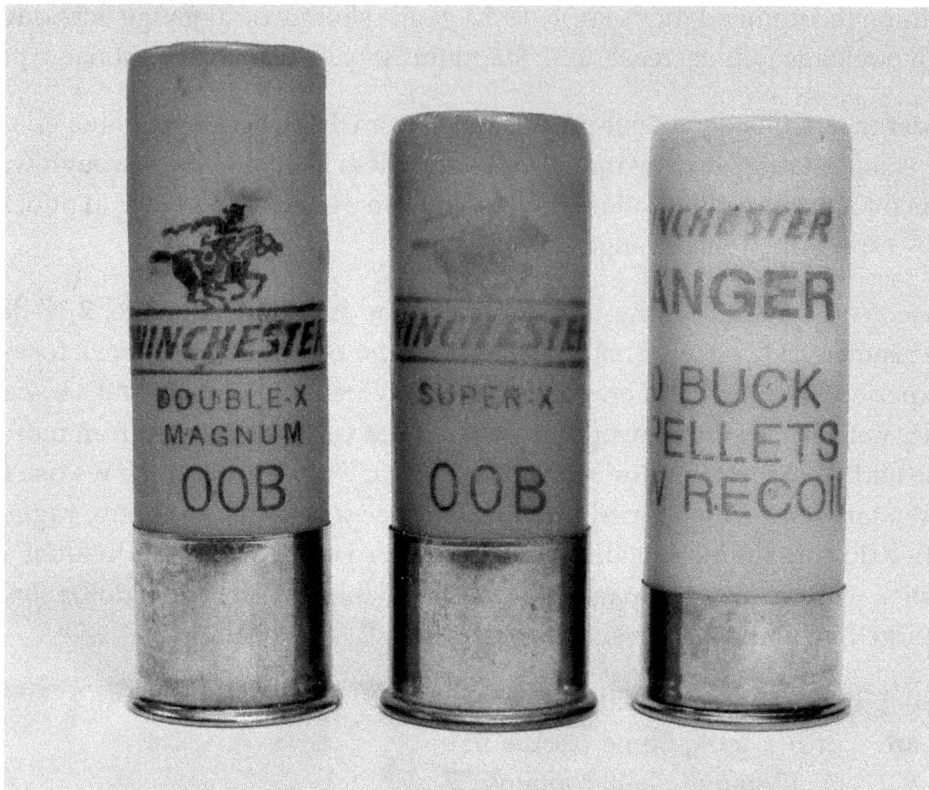

Three distinct 00 Buck power options with corresponding recoil. For many folks the low-recoil load will provide better overall control. Hunters may want more power.

The sealed-crimp buckshot shell (R) prevents leakage of buffering material inside magazines.

For civilians, obtaining the low recoil tactical loads may be a challenge. In that case, peruse the websites of ammo manufacturers and distributors to see what strikes your fancy. The best results often equate to higher prices, but buckshot isn't typically fired in large quantities. You might want a stiffer load anyway, since "low recoil" means lower velocity.

Extended range and big game. That extra rifle-sighted 21" slug barrel is a handy accessory. We've fired thousands of plain Winchester Super-X, 2 ¾ inch, 1-ounce slugs from our smoothbore barrels. Zeroed a couple inches high at 25 yards, shots will be on at 50 yards, and a few inches low at 75 yards. Things start unraveling there, drop and wind being significant factors. However, we've fired slugs out to 200 yards more than once using so-called "Kentucky windage" and holdover (meaning we guessed). On one occasion I was spotting for another instructor as he fired at steel, torso-sized silhouettes from this distance. Holding high, Mike lobbed in big .73 caliber, one-ounce 'bullets' that were clearly visible through my scope. They were fun to watch, and flattened themselves with very audible impacts, resembling wide, thin lead wafers after impact. Hits were surprisingly reliable. Admittedly, this was mostly just a stunt. Federal's "Truball" slug is supposedly fairly accurate in smoothbore shotgun barrels. We haven't fired a large enough quantity to establish the rumor, but we did see encouraging initial results.

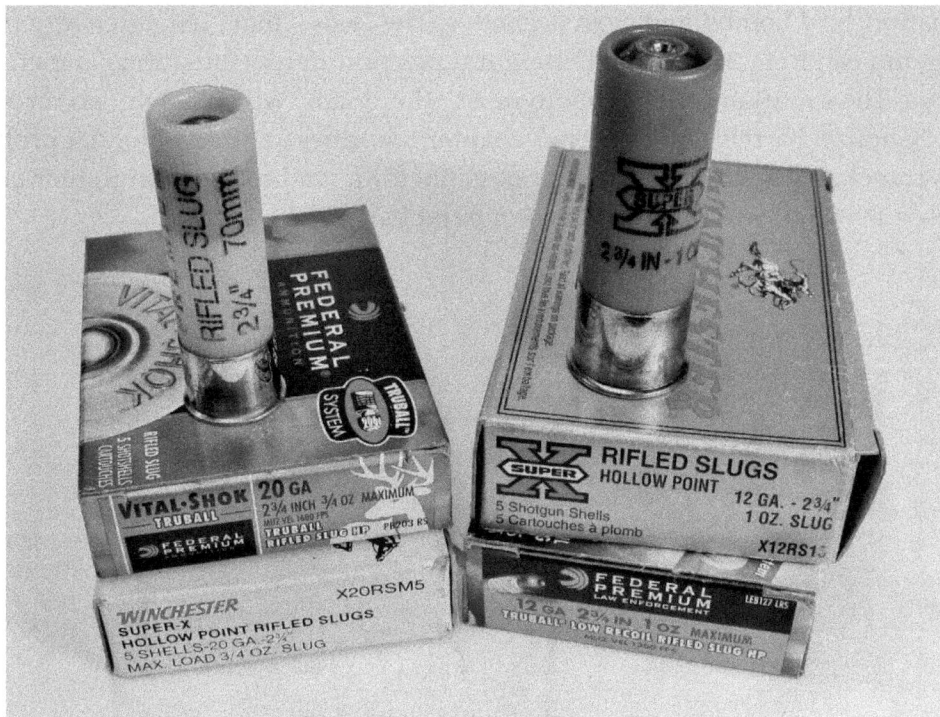

A useful assortment of 20 & 12-gauge slugs. The old
Super-X is our 12-gauge general-purpose load.

On the other hand, using newer sabot-clad projectiles designed for rifled barrels, a 200 yard shot is completely feasible with the right optics. Several major scope manufacturers offer shotgun-specific models. Reticles with calibrated holdover marks permit accurate sabot slug placement at extended ranges. Your shotgun becomes a serious rifle with the right combination of projectile, barrel, optics, and rangefinder. The only problem is that shot loads go haywire in rifled bores. That's why a dedicated gun or accessory barrel makes the most sense.

Staggered or alternate load strategies. We're familiar with the idea of combining buckshot and slug loads in the same magazine. It sounds reasonable in principle, but reality is another matter. Combining the "K.I.S.S." concept with Murphy's Law, the dual-load presents issues. This point is easily illustrated through drills involving multiple targets at various ranges. Those with average skills will quickly lose track, especially if sustained loading (a desirable action) occurs. In our experience, you're far better off segregating different loads. Even then, shell switches are dicey without well-grounded techniques.

The "Basic Load." This term refers to a planned supply of ammunition adequate to cover most circumstances. In this case, we're talking about a practical quantity for personal carry. My defensive selection includes 7 00 Buckshot shells in the magazine, 6 more in a bandoleer (or side saddle mount), plus two separate slugs. I can switch to those fairly quickly thanks to regular practice. Fifteen shells may not seem like many, but the weight adds up. You'll want more in reserve, because defensive ammunition should be rotated annually, or more often during harsh conditions.

Specialty loads. These include rubber bullets, bean bags, chemical agents, lock busters, flash bangs, blanks, illumination, bird bombs, and even flechettes. The "less lethal" kinetic energy rounds require proper training and can in fact be lethal. Different types fit different situations, depending on range and other factors. The same concerns apply to most other loads, which is why no recommendations are made. The "Chokes, Barrels and Patterns" chapter identifies another potential problem. Some of these products travel at low velocity and it is possible for a wad to lodge in tighter choke constrictions. Any subsequent shot will probably cause a burst barrel.

Two unusual 12-gauge munitions. The bird-bombs are used for pest control.

An assortment of 12 Ga. "less lethal" projectiles, including bean bags and rubber bullets. Many liabilities are attached to use of such munitions.

A word about corrosion. The so-called "brass" portion of shotgun shells may actually be mild steel with copper plating. They may rust over time, or if exposed to the elements. This causes concern during long-term storage. Clothe shell loops can also contribute to corroded bases over a period of time. Some rust preventatives (like WD-40) can kill primers. If a magnet sticks to the base, they're ferrous. Another concern involves steel shot, which can rust after exposure to water – especially salt water. The pellets can fuse together, acting like a heavy-duty bore obstruction when fired. Plan accordingly when acquiring a supply of shells.

Ammunition of all types does best if stored in a cool, dry place. GI ammo cans also do a great job. High humidity is bad, so sheds or damp basements are out. Prolonged exposure to high temperatures is also bad. Heat can deteriorate powder, so un-insulated attics and vehicles will eventually cause problems.

CHAPTER 9

CLEANING, MAINTENANCE, AND STORAGE

We're cleaning shotguns all the time. It's not rocket science. We'll often mass clean a rack full of M 870 range guns using a simple process. The techniques that follow are geared towards these guns, but probably will work with other models. Some of the gas-operated shotguns are a bother, which is why many folks ignore this chore and just fire them until they quit.

Safety. First, the obvious (we hope): Look the gun over to ensure it is unloaded! Then do it again. Pay special attention to the magazine follower, so it isn't confused with the base of a live shell. Move any live ammunition to some out-of-reach location, and put on a set of safety or shooting glasses for protection from solvent splatter or flying springs.

When the job is finished, wash up thoroughly. Lead exposure is no joke and, besides direct ingestion, residual lead can be inadvertently deposited on furniture, clothing, or vehicles. Foster-type lead slugs will quickly plate a shotgun barrel, and particles can fly in all directions while bore brushing. Shotgun cleaning is nothing you'll want to do at your kitchen table. It's better done outdoors, which is exactly what we prefer to do when performing basic field cleaning on a rack of dirty guns. More detailed disassembly, such as removal of trigger groups, is better done in a contained environment to aid in recovery of errant parts.

A tackle box makes a good service center. A workbench helps, too. Remove any live ammo first (the metallic cartridges are actually home-made dummies).

Solvents and lubricants. We use the following products and all but Break-Free are sold by Venco. Be sure to read their labels prior to application. Some readers will no doubt disagree with our choices, but after steady use for many years, we know they work.

➢ **Shooter's Choice:** This is our go-to solvent. It doesn't have the aroma of Hoppes, but it is hell on fouling of all kinds, including plastic. Those with camo-finish guns should probably apply a test drop to an inconspicuous area, just to make sure nothing dissolves. So far, we haven't seen damage to black, synthetic stocks, but we wipe them off immediately.

➢ **Break-Free CLP:** CLP stands for "cleaner, lubricant and preservative." We use it for general lubrication of metal-on-metal contact points. It also works well for a final wiping of metal surfaces, and it inhibits corrosion. I've seen controlled test results involving ferrous metal and salt baths. This product rates high. The military uses it for the same reason.

➢ **Gun Scrubber:** Sold in aerosol cans, this evaporative solvent is similar to brake cleaner. It cuts fouling in hard-to-reach areas, but will leave parts bone dry. Be careful not to get any in your eyes, and keep it away from night sights. It also kills hornets!

➢ **Rust Prevent:** As a final step, we often spray a film from an aerosol can. It smells good, too!

A few tried and true solvents. We buy Shooter's choice by the gallon!

One controversial product we use is WD-40. It can prevent surface rust, but I've heard very knowledgeable instructors say they'd rather use weasel pee on firearms. Arguments against it are that it eventually builds up a thick deposit, and that WD-40 will kill primers. Both are true so moderation is the key. If time is short and we have a rack of shotguns that have been exposed to rain, we'll often reach for a WD-40 can. We spray it on the metal surfaces and then wipe off the excess with a rag.

"WD" stands for water displacement, which it does well. We're careful not to get any on live ammo, or inside a magazine tube. I've used a lightly misted rag for years to wipe down personal firearms after handling and have no rust issues whatsoever. On the job, we only use it for triage during wet weather. A thorough cleaning immediately follows in a drier location.

Rods and brushes. You can spend a fair amount of money on a fully decked out cleaning kit. A few of the fancier ones include pretty wooden rods. Short thread adapters are available that allow conversion of thinner rifle rods to larger-diameter shotgun accessories. Flexible, pull through devices, like the "Bore Snake", have become popular, too. A weighted end is dropped into the open breech and down through the barrel. After it exits the muzzle it is pulled through from that end as bristle and swab sections do their magic. It'll get dirty, but can be washed. A flexible system is a good alternative for occasional use by those on the move, but it can't dislodge a stuck wad. We always have at least one rod nearby.

➢ **Rods:** Ours look more like battlefield artifacts. We actually use two aluminum rods without their handles. Both are jointed and, with shorter barrels, we remove a section for more convenient handling. The first holds a bore brush, and the second is for swabbing.

➢ **Brushes:** We use Brownell's "Double-Tuff" bronze bore brushes. When they wear enough to lose effectiveness, we use them as patch holders. Depending on their shape, they'll sometimes work in 20-gauge barrels, too!

➢ **Bore-swab:** A worn out brush makes an okay patch holder. Just wrap a patch around the end and the bristles will keep it in place. A nice alternate device is a Bore-Tech plastic jag, specifically designed to hold a patch. It has a pointed tip and flexible collet-like fingers, which maintain good bore contact. We don't use the slotted tips with loops in shotgun barrels, because they don't provide full bore contact.

➢ **Patches:** For 12-gauge cleaning, we use 3-inch square patches sold by Brownell's. They'll work with 20 Ga., too. In fact, we also use them for many centerfire barrel cleaning chores; especially those involving 9mm - .45 handguns. One difference: We use a slotted tip for proper fit.

Field cleaning techniques. Once a barrel is removed, it's easy to just shove a rod through it. Most pumps and autoloaders use their threaded magazine cap to secure the barrel, permitting disassembly. To remove the barrel from a Remington M 870, open the action part way until the forend sides conform with the barrel's taper. After the cap is unscrewed, the barrel should pull right off the end of the magazine tube. It's not a bad idea to then reinstall the

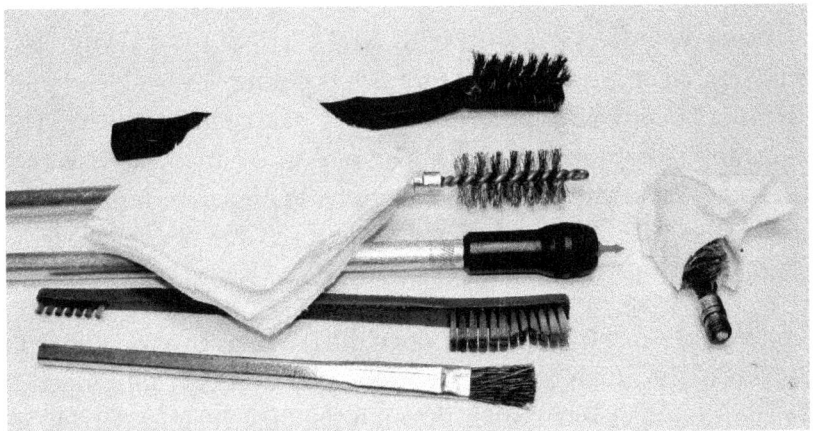

A few handy items. The pointed-tip rubber patch holder works well but a worn bore brush will do the job.

cap to prevent the large magazine spring and its retainer from launching with vigor. Those with extended magazines will probably need to remove a clamp that secures the extension to the barrel. At that point, the highly compressed spring will pop free with considerable force. You might as well pull it out and let the follower freefall into your hand.

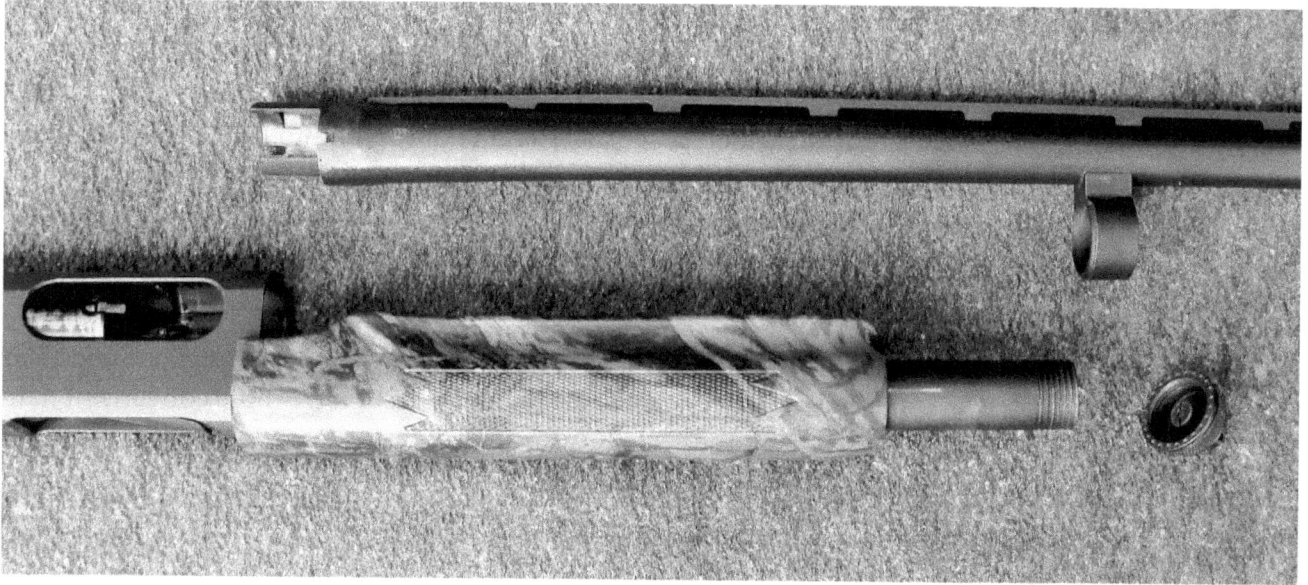

Remington M-870 barrel removal: The slide is only part way back for better clearance. Screw the barrel nut back on to maintain magazine spring capture.

If you clean regularly, a few passes with solvent and a brush should loosen up most of the crud in the barrel. Slug residue requires more elbow grease, and some shells also burn dirty, leaving plenty of powder residue. Removal of the barrel allows you to insert the rod through its breech, which is less tight than the muzzle/choke end. It's also easier on the brush. The rear of an M 870 barrel has a sharp protrusion, which can cut your hand as the rod pops through the muzzle. Without a handle, it's easy to apply pressure to the rear of the rod, using a hard surface like a bench top for the final push. Just pluck the rod out from the muzzle end afterward and repeat the process.

Because we often wind up with a rack full of fired shotguns with limited time, we may reverse the process, cleaning vertically through the muzzles with the guns still assembled. You can bet we clear them at least twice before doing so! A patch is first inserted across the opened bolt face to serve as a flag, and to capture any crud. The reason we resort to this is because our guns have extended magazines, which complicates disassembly. With a standard magazine cap, barrel removal is easy. *It's also a whole lot safer to entirely avoid the business end of your shotgun!* Cleaning through the muzzle is best avoided with any rifled barrel, since damage to its crown may occur.

With our two-rod technique, once the barrel is brushed out, it's easy to switch to the second one with its worn out brush or jag. Of course, you don't actually need the extra rod, you can just thread in the tip you need. A few passes through the bore from breech to muzzle should finish the job.

Choke tubes are unscrewed afterward and lubed. This step doesn't take long, and will prevent future headaches. A few drops of lubricant wiped into the threads will suffice. Don't run a patch into the barrel's threads since it can leave fibers that may bind the choke tube.

When cleaning guns like M 870s, after barrel removal, you can nudge the bolt forward until its face emerges from the receiver. Make sure the extractor moves freely and all fouling is cleared. Auto-loading shotguns will often have some sort of piston arrangement that also needs attention. They normally run best without oil. Apply a couple drops of lubricant to the operating rod or action bars and wipe down metal surfaces. Cycle the action a few times and you're done.

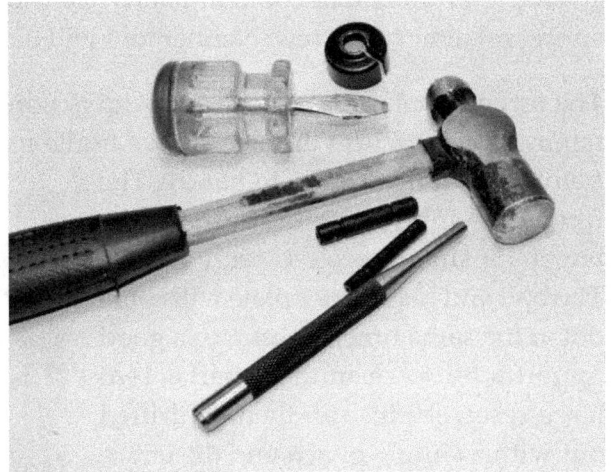

These basic tools will handle all basic maintenance chores except forend disassembly.

During routine cleaning you can nudge the bolt forward to brush off the bolt-face and extractor but, at some point, it'll need to come out.

Detailed cleaning. Magazine springs and followers need to be periodically removed so the tube's interior can be scrubbed. Failure to do so will result in a gradual accumulation of fouling which will eventually bind up a shell follower. Exposure to wet weather can produce similar effects caused by rust. Trigger groups should generally come out at the same time for cleaning. The simpler barrel cleaning only takes a few minutes, but if your gun is exposed to pouring rain, you'll need a more detailed regimen.

A couple of tools that come in handy for Model 870 owners are a forend wrench and a magazine spring retainer corkscrew. Neither tool will be used often, but both will be appreciated when needed.

The forend contains a steel sleeve with action bars that encircles the magazine tube. A threaded retaining collar secures the steel sleeve to the forend. This retainer can loosen after hard use, at which point the wrench comes in handy. The forend assembly must first be removed, but it's a simple process with an 870. The bolt and bolt carrier plate will come out at the same time, presenting a good opportunity to clean these parts. Two large receiver pins can then be drifted out with a simple punch and the entire trigger assembly can be removed for maintenance.

Depending on the gun, the magazine spring may be contained by a spring steel cap located under the barrel nut. It's a pain to pry out, which is why that corkscrew type wrench is handy. The tapered tip is threaded into the cap and pulled out. Lacking this tool, you can work it free with a screwdriver. In either case, be prepared to catch that magazine spring! Some manufacturers use other spring retaining designs, including Remington. Remember the two magazine tube indents that limit shell capacity? Well, they also engage a plastic retainer insert, which can be pushed inward and rotated 90 degrees for removal. As previously mentioned, magazine tube extensions normally serve as both barrel nuts and spring retainers. In all cases, once separated, remove the magazine spring and follower. Your gun may also have a "plug", which limits magazine capacity to 2 shells. It should drop out with the spring and follower. Wipe everything off using a patch dampened with Break-Free.

The magazine spring retaining collar can be coaxed out with a screwdriver if you hold your mouth just right. Be ready to capture the energetic magazine spring.

A M-870's bolt and forend assembly can be pulled forward and out while depressing the magazine's shell-stop.

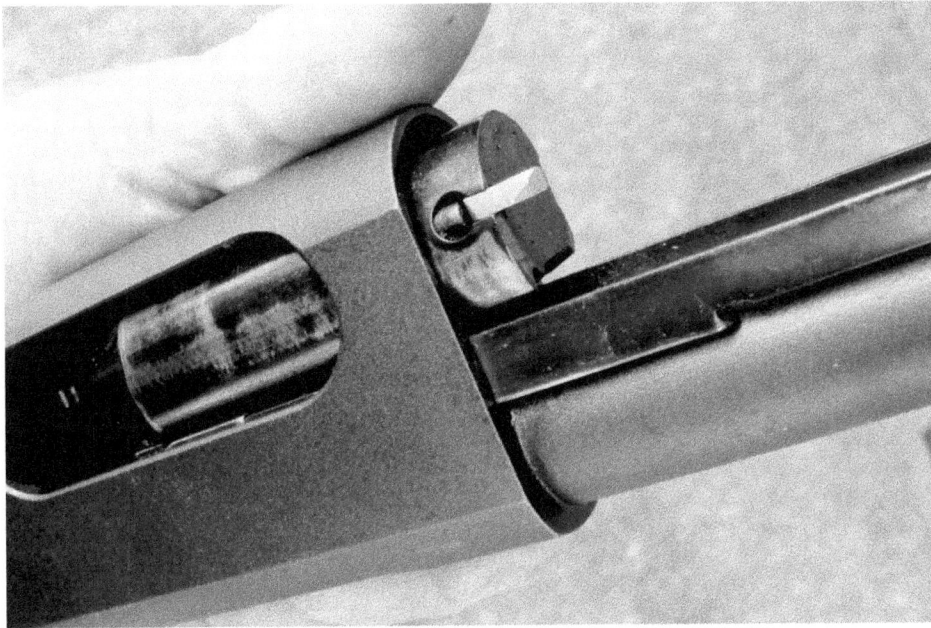

The bolt/forend assembly is now free and ready for removal.

The breech bolt and forend assembly can be removed as one unit. Retract the forend, reach into the bottom of the receiver, and depress the left shell stop. There are two located at the magazine's mouth. You want the port side stop, referencing off a shooting position. Holding the stop in with an index finger, the slide and bolt assembly can be pulled forward with your other hand, and then removed. The bolt sits on a flat carrier plate, which nests in cutouts on the two action bars. Upon removal, the bolt and carrier can fall free, so this chore is best performed over a soft surface. (A Remington M-1100/1187 semi-auto comes apart similarly, but the bolt handle first needs to be removed and should pull out with a firm tug.) Brush off any fouling on these parts, paying attention to the large, pivoting locking lug and extractor. Once everything is clean, you can place a drop of oil on the firing pin and bolt carrier hump which contacts the locking lug. By grasping the action bars you should be able to detect any movement of the steel sleeve inside the forend. Lacking a special forend wrench, a flat piece of steel plate can be used. It must first be ground to mate with the two recesses in the collar. It should be tightened while keeping the assembly indexed correctly in the forend.

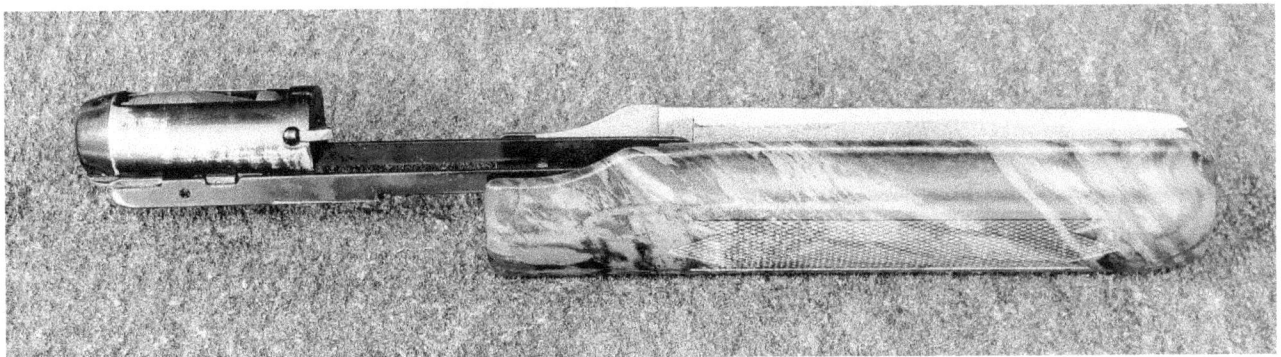

The bolt and carrier-plate, nested on the forend's action bars by gravity.

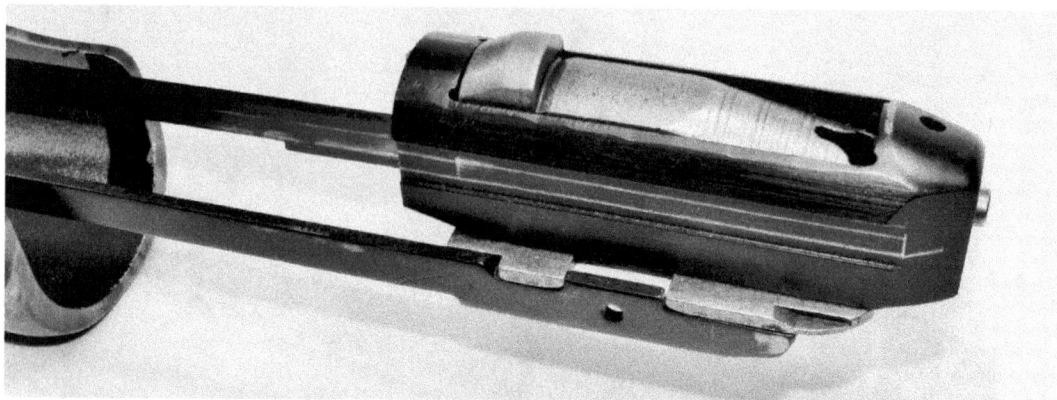

M-870 bolt with lug in locked position. The firing pin is visible on its rear.

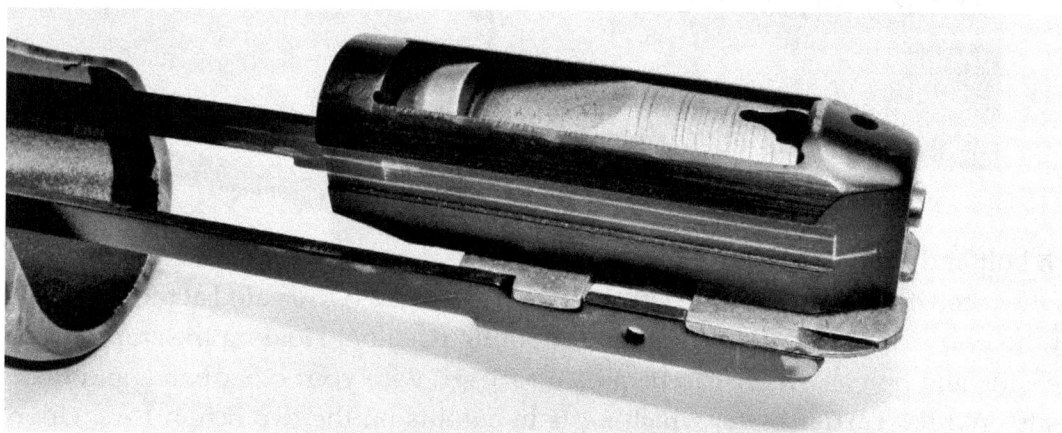

The same bolt in an unlocked position. Note its different location on the carrier.

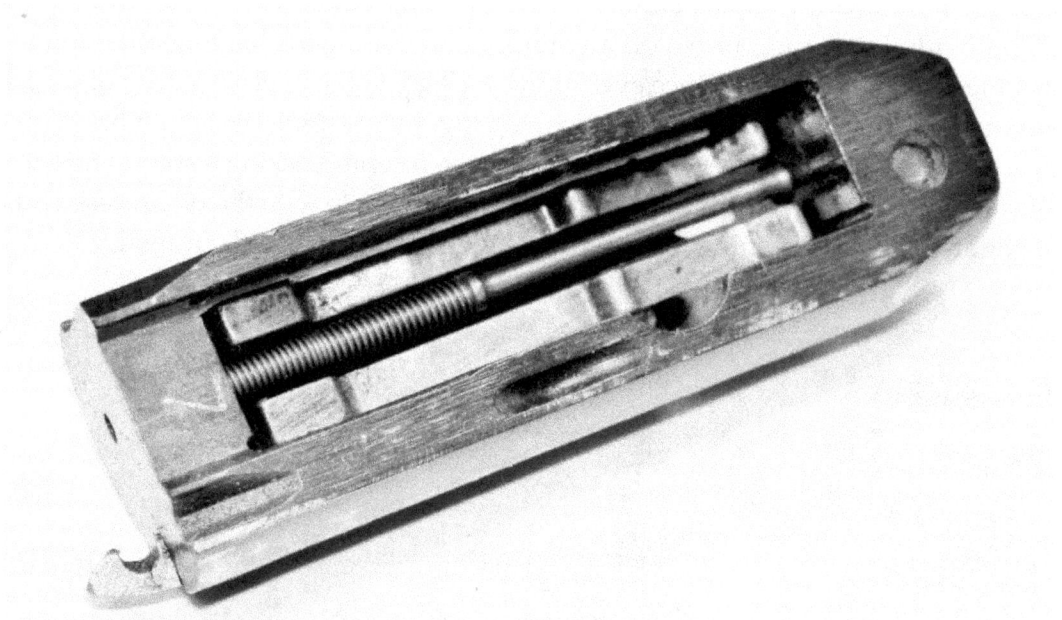

Bolt underside showing the extractor and firing pin. Apply a drop of oil to the latter.

M-870 bolt, with locking lug in "locked" position. Brush it off, along with the corresponding barrel mortise. Brush the extractor and apply a drop of oil to its plunger.

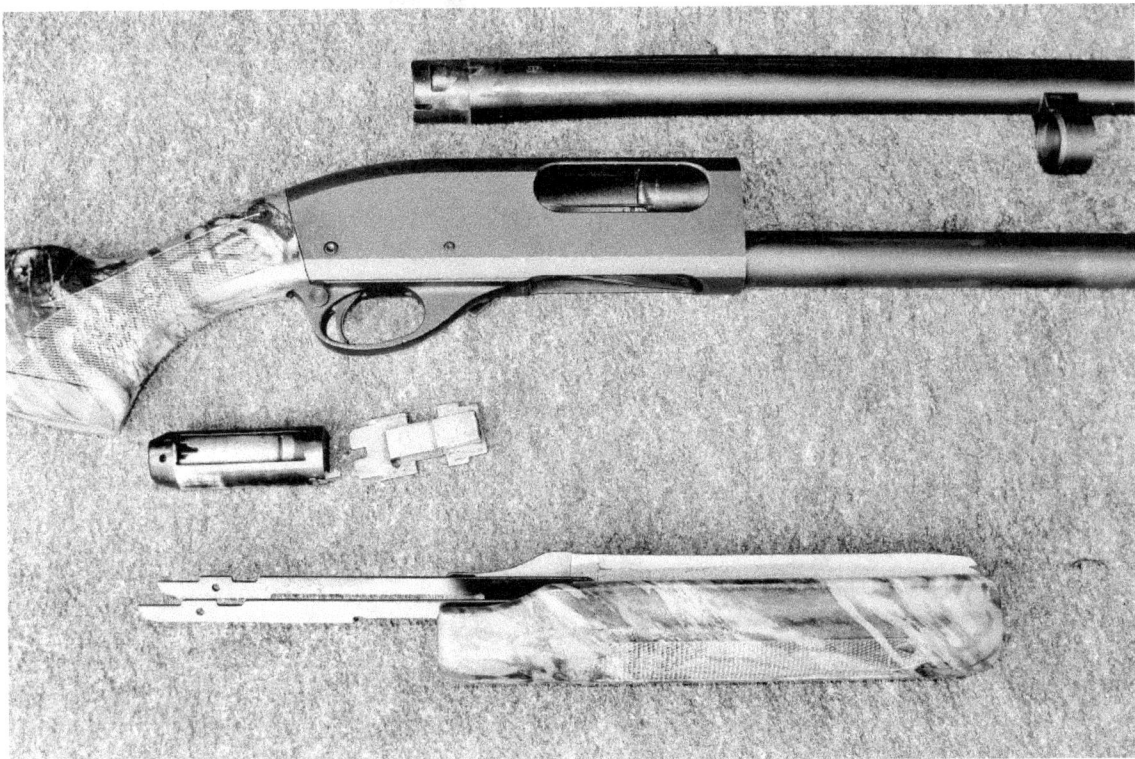

M-870 in basic field-strip mode. No tools are required with a standard magazine. Wipe off the bolt-carrier plate and action bars. Lightly lube with oil.

A look inside the forend, showing its retaining collar.

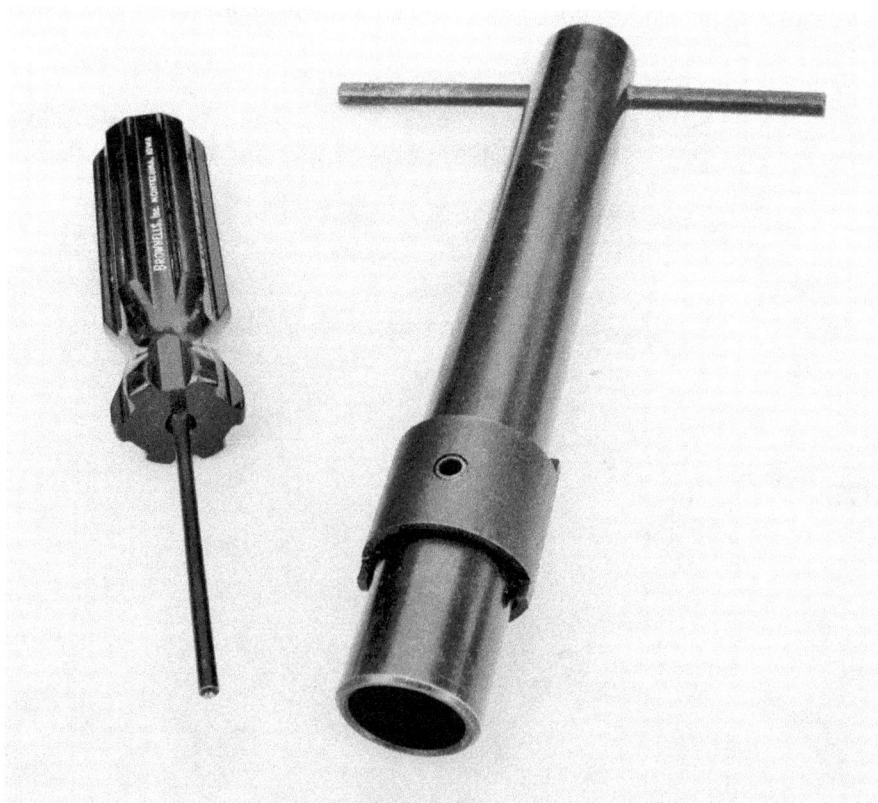

Two Model 870-specfic tools, a trigger pin punch and forend wrench. They're handy but not essential.

Two large pins retain the trigger assembly. They'll come out from either side, but we prefer left-to-right, which is common to many other types. We use a dedicated tool that resembles a screwdriver, with its tip ground to fit the concave ends of the pins. A punch will work just fine, but you may need a few careful hammer taps to drive them. I'm sure many folks just use a nail. Once the pins are removed, the entire trigger assembly will pull free from the bottom of the receiver. Carefully examine the relationship of parts in case anything falls out. This assembly is a good candidate for Gun Scrubber and is often full of unburned powder or debris. A few drops of oil on the moving parts will suffice. At this point, the inside of the receiver will be exposed. It may be full of junk, but at this point, cleaning is easy.

Driving out two pins will allow the trigger assembly to be pulled downward and out of a M-870's receiver.

M-870 trigger assembly, complete with a minor accumulation of debris. It needs periodic cleaning.

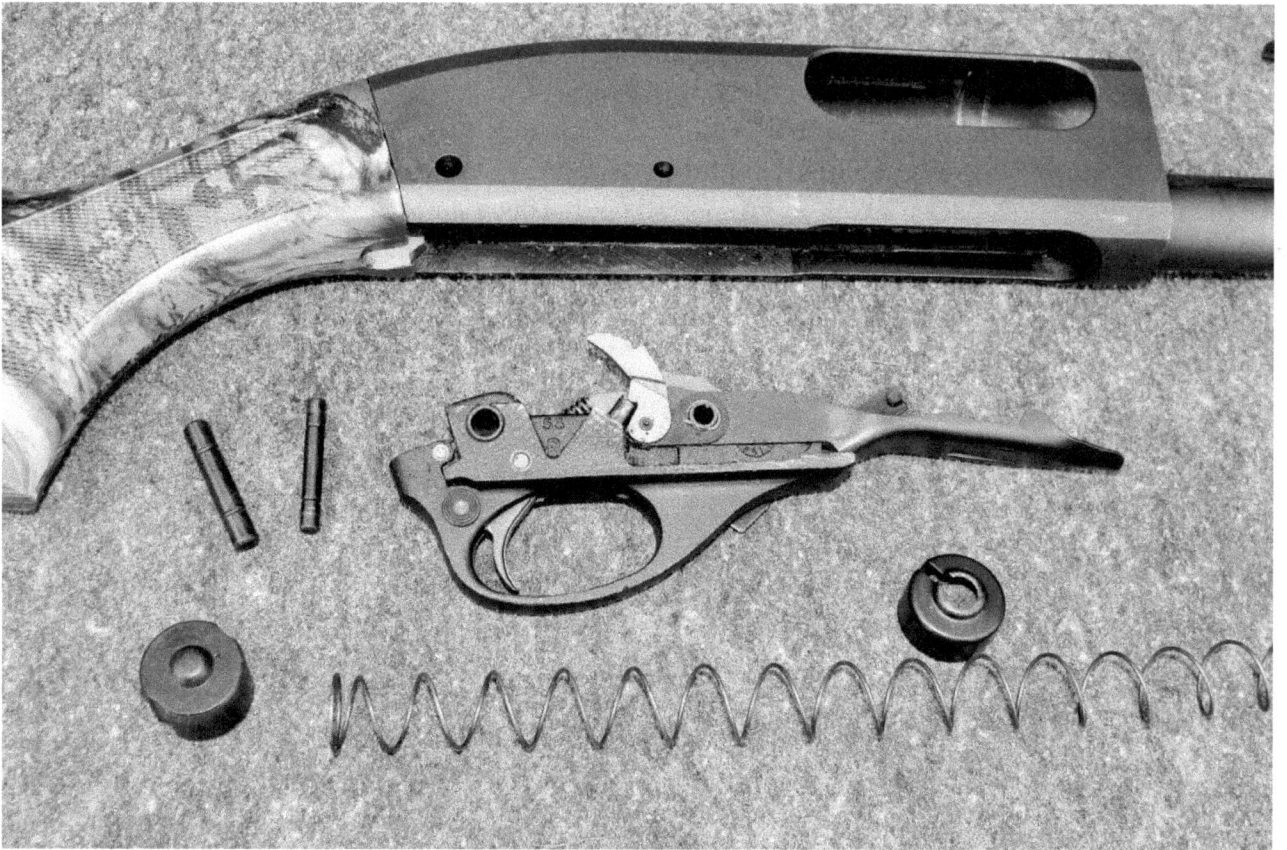

A stripped M-870 receiver with factory-issue magazine follower, spring and retainer. The spring will rust if exposed to water.

A M-870, user-disassembled for periodic maintenance. The process isn't difficult, and is worth your time after hard use.

The process for cleaning an 1100/1187 (and many other brands) is not much different, but a spring-loaded plunger is recessed into the rear of the receiver, and often accumulates dirt. It's actually the cap for the recoil spring and travels rearward into the stock. It needs to be clean so it has enough force to shove the bolt back into battery. A long, steel strut connects it to the bolt and its tip needs to nest in the plunger during reassembly. That can be a bit tricky, but it's possible with patience. The recoil spring should be replaced periodically depending on the quantity and types of shells fired. Heavy loads will exert more recoil, but even a steady diet of light loads can take their toll. Long-term storage with the bolt locked open just makes things worse. After a good cleaning, if the bolt seems sluggish when going into battery, it's probably time for a new recoil spring. For many people, replacement may best be done by a gunsmith. Most semi-autos will need to have their stocks removed in order to access the tubular housing. I change my own with caution, recognizing that a misstep can send a screwdriver tip through the side of a wooden stock.

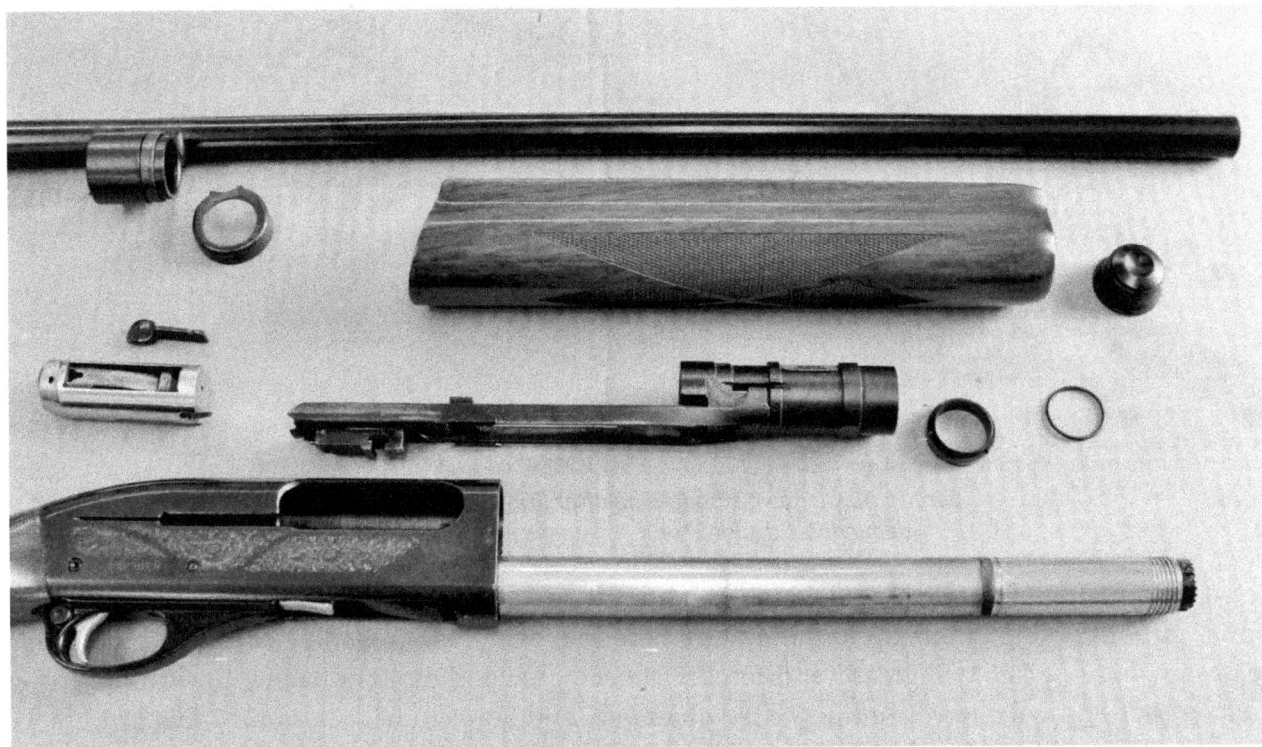

A stripped Remington M-1187. Note the carbon fouling on the magazine tube.
The O-ring slips into the annular recess. A spare is worthwhile.

One neat tip we picked up from a Benelli factory rep involved storage of an inertial gun. He recommended muzzle-down rack placement, which prevents lubricant from accumulating around the recoil spring plunger. Supposedly, it can act as a hydraulic dampener, slowing bolt speed. Although Benelli recommends 3 dram minimum loads, lighter shells may function when this trick is used. It's an old idea that's been used for decades to help keep oil from leeching into the wood stock heads of double-guns. A thick layer of lubricant in that area attracts all sorts of debris that can affect function. I always disassemble hunting guns at the end of each season or, sometimes, more often. It's surprising to see what lurks inside the action; weed seeds and pine needles are the usual suspects, as well as unburned powder flakes. The receiver's bolt slot makes a great pathway for small, foreign objects. The trigger housing is right below it, and shouldn't be ignored.

Beretta AL390 gas piston fouling after firing 100 shells. It's a dependable design, but it still needs maintenance!

We've seen plenty of autoloaders with sluggish bolts and most were the result of neglect. Besides debris, carbon fouling builds up to the point where they won't run reliably, causing a bum rap for the design. Lack of care will greatly shorten the life of a gun, and may contribute towards corrosion. Many higher-end guns have chrome-plated chambers and bores to help minimize this process, but rust can still develop – sometimes in a surprisingly short period of time. A pitted chamber will cause extraction problems as fired shells swell into its walls. More than once, I've seen a fine layer of rust appear within a few days of firing. Humidity probably doesn't help. You can save a gun by getting on this situation quickly using 0000 steel wool wrapped around a bore brush and impregnated with WD-40. Give the accumulation a good scrubbing and wipe it out with a clean patch.

STORAGE

Periodic inspections and basic cleanings go a long way toward peace of mind and will help maintain your investment. The means of storage is equally important. External corrosion can appear in a surprisingly short amount of time and if undiscovered, can cause pitting and permanent damage. Long-term gun storage in a case may transfer moisture from padded surfaces, which can accumulate humidity. We've seen firearms literally covered in water droplets after their soft cases were exposed to intense sunlight. These guns were properly stored at home and cased for a trip to the range. The soft interiors had probably absorbed moisture over time and placed in the sun upon arrival. Their interiors became saunas, causing "sweating" that condensed on the guns. It happened to me and a fast wipe down saved the day. The case was spread open to dry in the sun. If for some reason the gun had remained in the case, things would have turned out differently.

The same situation can occur when entering a warm area from low temperatures. Moisture will quickly condense on a cold firearm, covering it with water droplets. Those spending time in remote winter locations often leave their working gun outdoors for this reason. A gradual warm up with constant attention is otherwise necessary.

Besides gun cases, wrapping a gun in cloth can wreak havoc, even if stored indoors. Any damp place like a basement or shed is just as bad, but long-term vehicle storage can cause similar problems. Ammunition is also best stored in a cool, dry location. It will last for decades with care, but heat will cause rapid deterioration. Moisture is as bad for ammo as it is for guns!

That's why a good gun safe located in a dry and stable environment is worthwhile. Desiccant packs and dehumidifier rods should provide adequate protection from rust if some thought goes into the final location. I haven't had any issues using a safe alone, which is located in a primary living space. The extra peace of mind afforded by a good safe is another major dividend. *Survival Guns: A Beginner's Guide* covers safes and other storage options in detail. We all have an obligation to ensure that our firearms and ammunition don't fall into the wrong hands. A locked case or trigger lock may deter kids or honest folks but are really just interim steps.

Springs and storage. We prefer to relax the tension on a few key springs during long-term storage. With a pump gun, the hammer and magazine springs undergo significant compression. Opening the action will cock the internal hammer, which is typically powered by a strong coil spring. A fully loaded tubular magazine compresses the long spring that drives its follower. A semi-auto also has a strong recoil spring, which undergoes compression while its bolt is locked rearward. On a range, it should be placed in a vertical rack that way for safety's sake. It'll be fine for the short-term, but don't store a semi-auto on a long-term basis in that way. A weakened recoil spring will degrade function and could cause damage to the gun from excessive bolt velocity.

Our armed agency has a series of guns always loaded to "carry condition." Their chambers are empty, but the hammers are cocked (which locks their actions closed), and their magazines are fully loaded. So each year we rotate our inventory of shotguns *and* ammunition. Those guns in active service are pulled for storage after a detailed cleaning procedure. Another batch of guns then steps up for an annual tour of duty with fresh shells. The guns relegated to storage remain unloaded and un-cocked (we just dry fire them). Since we have enough guns to go through three cycles, within a three year period each gun's springs are only fully compressed for one third of that time. Does it matter? Truthfully, we're not totally sure. We have a series of M 870s that are more than 20 years old and each still has its original hammer spring. This translates to more than seven years of regular compression. So far, we haven't had any misfires or light strikes either (our annual ammo exchange no doubt helps). On the other hand, we have replaced a few magazine springs. They still worked, but seemed a bit less energetic. And in real life, are you going to buy three identical guns and rotate each one annually? Instead, you might want to buy a spare magazine spring, easily swapped out. After a few years, consider gunsmith servicing of your trigger group with a hammer spring replacement. If you have a semi-auto, the recoil spring can be replaced as well.

Final storage thoughts. I don't keep a loaded shotgun at home (in the upcoming training chapter you can learn how to load one quickly). So after verifying that a gun is empty, I just dry fire it. That's pulling the trigger on an empty chamber, which drops the hammer and de-compresses its spring. You won't hurt an M 870, but some guns can be damaged by regular dry firing. I use snap caps to un-cock any double-guns. They are special dummy shells with spring-loaded "primers" to cushion each hammer blow.

The two "shells" are actually snap-caps, designed for relieving hammer spring tension during long-term storage. They can be easily identified once in hand.

Sources. Brownell's was listed in the accessories chapter. Their original market was gunsmithing supplies, which is still a large part of their business. Cleaning and maintenance go hand-in-hand, so you should be able to locate all of the essentials, as well as plenty of other trinkets. Midway USA also catalogs a nice selection of gear.

CHAPTER 10

SAFETY AND TRAINING

Some people think you merely need to point a scattergun in the direction of a target to make a hit. (Not true!) Some fear recoil, some even expecting to be knocked right off their feet. Actually, I have been tipped over on a few occasions, due entirely to poor balance. Most recently, I went over backwards while seated on a pair of stacked milk crates in a blind. The target was a high, overhead crow that passed to my rear. I wound up leaning too far back and keeled over in some bushes upon the shot. My sidekick thought it was pretty funny, and things turned out fine (the crow would disagree), but the situation could have been perilous if my finger remained in the trigger guard. Since I was shooting a semi-auto, the gun probably had a new shell chambered before I hit the ground. With a pump, this danger would have been eliminated.

Small-statured shooters and big, hard-kicking guns don't mix. Some adults find entertainment in subjecting unprepared shooters to heavy recoil and that's just stupid. In some cases, loss of footing can occur. You can also plan on a new shooter having a nearly incurable flinch. On the other hand, with a bit of planning, most folks can handle recoil just fine. Any average adult should be able to manage the recoil of a 12-gauge with light shells. As covered previously, this combination in a gas-operated semi-auto will provide a soft-shooting gun, which can be initially loaded with only one shell. But, just about any gun of average weight will be mild with the lightest loads and a pump gun will feed them. In some cases, a youth-sized 20-gauge is the logical solution. That too can be loaded with only one shell; not a bad way to start a new shooter.

When it comes to claybird lessons, I start small shooters on a 20-gauge gas-operated semi-auto. I use a Beretta Model 391 youth model with a 24-inch barrel and shorter stock. Recoil is mild and the smaller dimensions ensure a proper fit in most cases. We always start by loading only one shell. The bolt locks back after the shot, so the action is automatically opened. Frequent loading reinforces correct operation, and the pause between loads allows time to analyze technique.

Beretta M-391 20 Ga. youth gun. Less expensive options exist.

Once properly equipped with the right gun and shells, the following exercises and live-fire drills may prove useful. The targets are stationary and geared toward defense. We can think about integrating these defensive techniques with hunting-oriented challenges such as clay pigeons later.

Although geared toward a Model 870, many of the basic principles we'll cover work with other systems. Before getting started, let's think about a safe learning environment.

SAFETY FIRST

You can cause a whole lot of damage with any firearm, but a close range hit from a shotgun will be devastating. A load of birdshot fired into a paper combat silhouette target from 5-7 yards succinctly makes the point! Complete awareness of your muzzle's direction is necessary at all times. Proper handling requires your total attention and unwavering adherence to basic firearm safety rules. Don't attempt to fire a shot until these five "Universal Firearms Safety Rules" can be recited by heart.

#1: Treat all firearms as if loaded! We don't care if the safety is on or not, or whether the gun is unloaded. It doesn't matter if the action is open or closed. It's *ALWAYS* loaded. Never sweep the muzzle by any part of you - or anyone else!

#2: Don't allow the muzzle to sweep something you're unwilling to destroy! This rule *always* applies, whether during handling or in storage. While participating in organized programs with pumps or autoloaders, unless engaged in firing, we teach shooters to utilize "safe range carry." The gun is unloaded and on "safe" with the action open. It is carried vertically, with its muzzle higher than any shooters' head. This is carry provides a safe start for new shooters. An alternate sporting carry for a break-barrel gun has the action open and muzzle down. We can live with this carry using other types, but please don't rest a barrel on your shoe. It's a bad habit that may cost you a foot! On a range, vertical racks are useful. Stow your gun with its action open, and don't grab the forend of a pump when you retrieve it. Further caution is necessary when using a sling. If carried muzzle-up, it's easy to sweep others while bending forward, and a careless rearward dismount can produce the same effect. Be aware of your muzzle at all times!

#3: Keep your finger off the trigger until ready to fire! Find an index point on the gun. Typically, the trigger finger will be extended forward, parallel with the bottom of the receiver. You can disengage a safety while committing to fire, but the trigger is off limits until the gun is fully mounted. You can bet that most "accidental" shootings are the result of failing to follow this rule.

#4: Be sure of your target, and what is beyond it! The smaller birdshot pellets have a fairly short maximum carrying range of less than a quarter mile. Buckshot and slugs will travel further, but all are devastating at close range. Multiple projectiles introduce further bystander concerns due to expanding patterns. We need to exercise situational awareness at all times and manage our muzzles accordingly.

#5: Check your shotgun upon handling it. Double barreled or break-action guns are easy to inspect since their chambers are exposed when the action is open. Autoloaders and pumps are a different matter. Tubular magazines can be hard to see, so check one and check it again. With the action open, you can use a finger for tactile verification. When handling an "unloaded" Model 870, push the carrier up after opening its action. If, for some reason, a stray shell popped out of the magazine, it should be sitting on the carrier where you can see or feel it. The magazine is easier to check as well. Be extra careful when changing choke tubes, performing that step with the action open.

Other concerns. It goes without saying that alcohol and drugs don't mix with firearms. Neither does unsupervised children or others raising safety concerns. Unattended firearms are a huge problem, requiring responsible storage. Firearms safety and storage is covered in-depth in *Survival Guns: A Beginner's Guide*.

A shotgun has a large muzzle, which can pick up foreign objects. Mud or snow will elevate pressures to unsafe levels resulting in a blown barrel. An under-powered shell can cause the same problem if its wad doesn't exit the bore. If you detect a mild report, stop shooting, unload, leave the action open, and check for an obstruction. It's not even a bad idea to do so if a gun has been in storage for a while.

Separate any shells of different gauges. The last thing (maybe literally) you'll need is a 20-gauge shell lodged in a 12-gauge bore.

Before firing a shot, track down a good set of shooting glasses and some hearing protection. Wear them anytime firearms are discharged in your proximity and make sure everyone else is properly equipped. Although not commonly known, small birdshot pellets can sometimes ricochet off trees, knock down targets, or even clay pigeons. Why risk losing an eye?

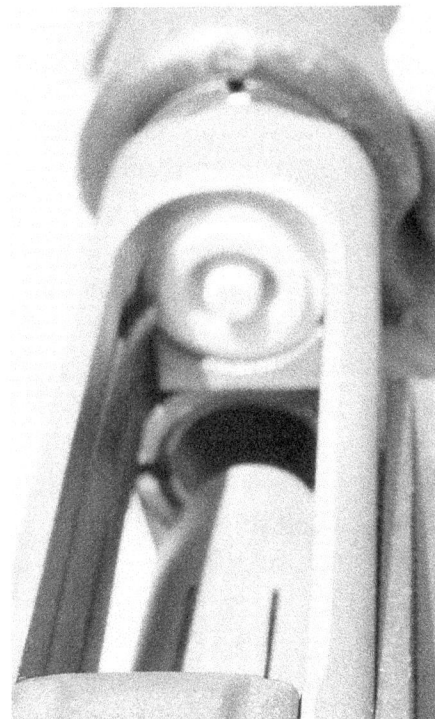

Upside-down view of a cleared M-870. The LE forend makes this somewhat easier.

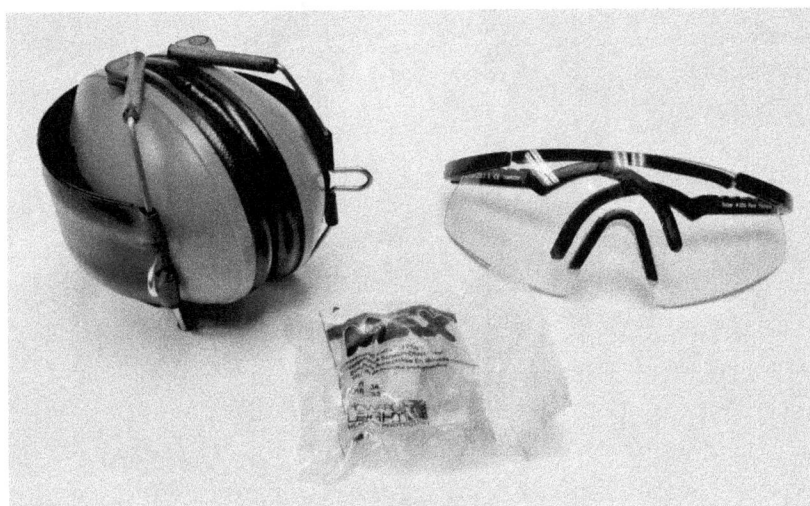

Basic hearing and eye protection: must-have items!

THE FUNDAMENTALS OF SHOTGUN MARKSMANSHIP

Five basic principles are often assigned to the process of firing a shot. Most often they are taught to rifle or handgun shooters, but they also extend to shotgun users. It's worth developing a stabile platform before attempting to fire your gun. Rather than delve into the tactical aspects used in conjunction with heavy body armor, let's keep things simple for now, employing techniques well-suited to manage recoil. We'll assume our shooter has a Remington Model 870 pump, and is also right-handed. Just reverse everything if left-handed.

Stance. Start with your left foot forward and your body turned somewhat to your right. Your feet should be planted shoulder-width apart, facing the same direction as your shoulders. Place a bit more weight on your forward (left) foot than your trailing foot. By then leaning slightly forward at the waist, recoil will transfer some weight rearward. Balance will be maintained by the trailing foot, and your body will act like a shock absorber, moving to soften the blow. When done properly, nothing is exaggerated and the body can serve as a stable but flexible platform. As for hitting fast crossing targets, there is more to footwork than meets the eye. The process is greatly simplified here to serve as a starting point. A good stance will permit flexibility to engage targets from multiple angles. Very wide foot placement can restrict the hips, limiting one's swing and causing poor follow-through. When engaging aerial targets, stance and swing are strongly connected as one smooth but dynamic event.

Grip. The gun should be pulled firmly into the shoulder socket by the shooting hand. It's not a death grip, but a solid grasp. The thumb should encircle the shotgun's grip to increase control while clearing the nose. The support hand grasps the forend with neutral pressure. We don't want to lose control during recoil or

A solid shooting position, capable of supporting controlled follow-up shots.

the gun may jump upward. Choppy cycling of a pump gun will also result, which can cause malfunctions. On the other hand, too much tension can cause other problems. If the forend is pulled forcibly rearward prior to shooting, the action may bind until the slide is tweaked forward. When executed correctly, everything should feel smooth but solid. If you extend the index finger of the support hand forward, it can serve as a subconscious pointer, which helps on moving targets.

The support-hand index finger can serve as a pointer.

Breath control. Although heavily reinforced during rifle and handgun shooting, not much attention is placed on this fundamental with shotguns. The former disciplines require deliberate aiming and precise bullet placement. On the other hand, a shotgun is most often fired reactively while pointed toward moving targets. One place where all disciplines merge is with the use of shotgun slugs. For all intents and purposes, the shotgun then becomes a rifle. Breathing interrupts the stability needed to fire an accurate shot. You can draw a few deep breaths and either hold it or gradually exhale while making a shot. Try to fire within 8 seconds or your vision will deteriorate. Where time permits, re-oxygenate your blood by repeating the process.

Sight alignment. This skill applies to iron sight slug barrels equipped with a rear and front sight. Once adjusted properly, reasonable accuracy is possible if both sights are correctly aligned. The shooter locates a front bead in a rear notch and then applies the image to the target. The front sight should remain in clear focus while trigger pressure is applied. Precise delivery of a single projectile is then possible. If a shot charge is fired through the same barrel, the broad pattern is more forgiving so the shooter can look through the sights, which only serve as a rough guide to direct a quicker hit. A bead-sighted bird barrel is designed primarily for use with shot loads, and is pointed rather than aimed. The process is very much like pointing your finger at a moving object, with focus directed to the target. The shooter's head and eye constitute the rear sight, so consistent head placement is necessary. Although some guns have a small extra bead serving as a rough rear sight, its main purpose is to promote centered head placement. The shooter's head should remain in contact with the stock before, during, and after the shot.

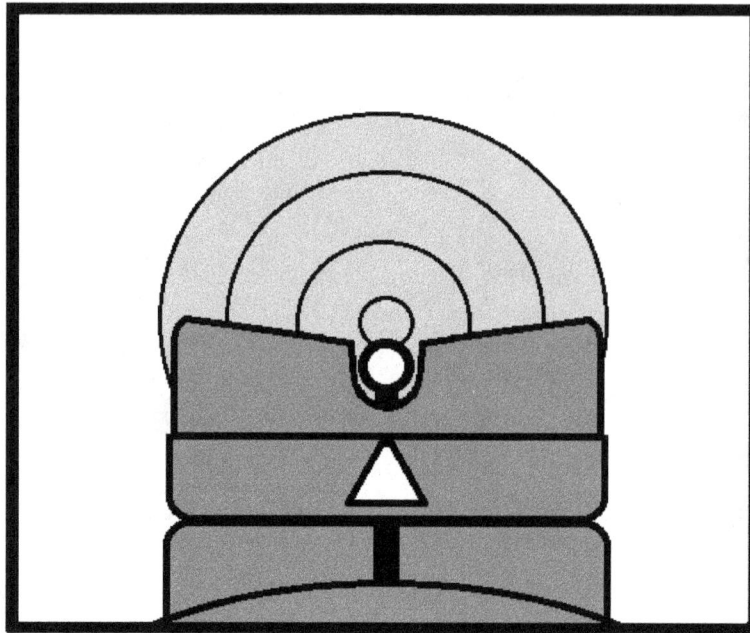

Remington M-870 sight picture using a slug barrel equipped with rifle sights.

Trigger control. Rifle shooters will be told to squeeze their triggers until the shot breaks as a surprise. The same principle extends to shotgun slugs. We gradually load pressure on the trigger while focusing on our front sight. At some point, the shot will occur but the exact instant will be unknown. The urge to flinch or anticipate recoil can thus be overcome, and an accurate hit will result. Fast moving targets are typically engaged with shot loads, providing little time for slow trigger finesse. Still, the trigger should not be yanked. Instead, it's a strong but controlled pull straight to the rear.

Summary. We may employ two techniques when shooting a shotgun. The first is very much like rifle marksmanship with careful sight alignment and trigger squeeze to permit accurate placement of slugs. The second is for shot loads and boils down to this: Stare at the target, keep your head on the stock, and pull the trigger. Either technique is much more effective when supported by a well-balanced position. Unfortunately, this may be impossible to achieve with a poor-fitting gun. Trying to start a small shooter with an adult-sized gun won't work. Balance and safety will be compromised. The gun will be too heavy and every step will be a struggle. Please consider this before proceeding to the next levels.

NON-FIRING EXERCISES

With a basic foundation established, it's time for some gun handling exercises. You can practice these in many places after ensuring your shotgun is unloaded. Chose a location with a bit of room and something to serve as a stand-in "target" (could be something as simple as a lamp). A few dummy shells are handy to have. They provide the means to gain familiarity with your gun without risking an uncontrolled discharge.

Dummy training shells. We use "action proving dummies", inert shells without propellant or a primer. They're not cheap, but they handle like live shells. We've used hand-loaded dummies with reservations: To ensure we're not handling a dud, the dummies are loaded with their primer pockets empty. Still, they could be confused with live shells so any live ammo must be removed from the training area! We're not too wild about light plastic dummies because they handle differently and won't stand up to frequent training activities. Snap caps are in the same boat, except for the solid aluminum "A-Zooms" sold in pairs. Five will get you by. You can dry fire a shotgun on them, too. Don't pull the trigger on any action-proving dummies with solid, non-primer pocket bases. Repeat firing pin strikes will dish in the bases and deform their rims, causing feeding and extraction problems.

Inert action-proving shells: our preferred training dummies.

*Other 12 Ga. types. Note the dented and dished action-proving dummy,
and live shell on right. Remove it prior to training!*

Note the home-made 20 Ga. dummy, compared to a live shell.

Note: *You can use live shells, but only in a safe environment! This will be a range for most people. Look your gun over carefully before proceeding, and consider the locations of all others!*

For now, shells won't be necessary. We'll just need to focus on a few basic positions...

Utilize effective "ready" positions. The transition to a proper shooting position will be much smoother from some sort of pre-fire "ready." Two useful versions are the "high ready" and "low ready." They can serve as starting points from which to efficiently mount the gun.

A classic high ready position.

➢ **High ready:** This position has dual-use advantages, for sporting purposes or defense. If done properly, you won't look like Cool Hand Luke with your muzzle pointed skyward. It looks spiffy on camera, but a more direct process will result from a straighter line to a target. In our "high ready" position, the stock hovers around the bottom of the ribcage while the muzzle indexes on the target. A taught string extending from the shooter's strong-side eye to the target would contact the muzzle or front sight. The business end of the gun isn't too high, nor is it too low. Think "eyes-muzzle- target." Since you're only moving the rear half of the gun, it's very fast. If done properly, the muzzle is sort of a hinge, remaining on the same plane, as the stock is moved smoothly to the shoulder and face. A few negatives: The high ready is a poor choice for entering doorways, and an unsafe position in a column formation. It's also tiring with a fully loaded gun. However, an unintentional discharge will send a shot pattern above a target, not into it. Admittedly, this could be a thrilling event associated with new underwear for all involved parties. Placement of the trigger finger outside the trigger guard is therefore necessary. The trigger finger can be lightly positioned against the safety button of an M 870, and it's reversible for right or left-handed shooters. It only takes a split-second to push it off during a gun mount.

➤ **Low ready:** This position is primarily defensive and can best be understood by watching coalition forces during maneuvers. The stock is on your shoulder and the muzzle is depressed. The degree of depression will depend upon the situation and proximity of threats. You don't want your muzzle to sweep your feet! For a generic low ready, imagine standing in a grounded hula hoop that contains your feet and muzzle. Acquiring a gun mount first will ensure that it returns to a correct firing position. The toe of stock acts like a hinge against the shoulder, so this position needs to be maintained for effective engagement. Smooth elevation of the muzzle will put you on target. Again, the trigger finger is outside the trigger guard. An unintentional discharge from a low ready can be frightening, or depending on the ground and angle, worse. Severe angles and dirt will result in lots of flying debris, but shallower angles and hard surfaces can skip-shoot projectiles with high velocity. Great caution is necessary, but it's a better position for addressing entrances, or during movement of multiple shooters. We see some people attempt to use the low ready on flying targets, but it's not recommended. For starters, you can sweep a bird dog during an exciting game bird flush. Next, you'll be introducing excessive muzzle movement while trying to lock on a target. If you watch the best claybird shooters, you'll see very little gun movement before and during the shot. What little there is will be smooth and efficient, developing from some version of the high ready. In other words, the low ready is better in serious social situations.

The low ready position.

Either position has advantages, depending on the situation. Don't discount the real advantages of an effective ready position. And never forget trigger finger discipline!

Develop an effective gun mount. Practice a smooth presentation to your shoulder and face. Don't bring your head to the gun. Instead, look at where you want a pattern to strike and mount the stock to your cheek. The butt should be on your shoulder, never your bicep. Although it's not tactical by today's standards, keeping your elbow parallel with the ground can help maintain the right position during and AFTER the shot. Bruises on your bicep result from the gun leaving your shoulder. Dropping your shooting arm can cause the stock to slide during recoil, especially while cycling a pump gun. With the elbow up, a pocket is formed to prevent this. However, a stock that is too long can cause similar problems by interfering with a proper mount.

A solid gun mount, necessary for effective repeat shots.

Eye dominance can also cause problems with proper head position. Ideally, the shooter's cheek will maintain contact with the stock, placing the gun-side eye directly behind the barrel. Roughly 80% of men have a dominant eye matching their strong hand. Women and kids are less predictable. If the off-eye is dominant, contorted head placement is likely. Fixes include squinting the cross-dominant eye, placing a small piece of Scotch Tape on the shooting glass lens, or switching hands entirely. That's often difficult with any adults who have previous shooting experience, but it can be an instant salvation for youngsters. For those without issues, practice shooting with both eyes open. Depth perception and peripheral awareness will improve, along with low-light vision.

Another great gun mount that positions the shooter's eye directly behind this set of ghost-ring sights.

Again, the cheek should maintain stock contact before, during, and after a shot. A high-shoulder gun mount helps. If the heel of the stock is positioned near the top of the shoulder, a fast, consistent mount will happen with less likelihood of your head coming off the stock. Failure to maintain contact will cause high shots. If a bit of the stock is actually higher than your shoulder, that's not necessarily a bad thing. This style helps prevent a rap in the nose from the strong thumb during recoil. With a well-fitted gun, keeping the shooting eye behind the barrel goes a long way toward consistent hits. With a rifle-sight equipped slug barrel, the same position should locate your eye for instant and maintained sight acquisition.

Mount and dry fire. Begin with the shotgun in "high ready." Execute a few smooth gun mounts before dry firing a shot. After things feel right, you can mount and dry fire a shot at a pre-designated target of reasonable size. It helps to try some daily dry firing until things feel right. At that point, you can pick up the tempo until you can dry fire a controlled shot when the gun contacts your face. A reasonable goal is to be able to mount and dry fire a controlled shot on a designated, stationary target within around 1 ½ seconds. After things feel smooth you can fool around with seasonal clothing. A paper plate located 12 yards away makes a good target for these drills.

Shoot-pump-watch dry fire. One more time: the gun should stay on your face before, during, and after each shot! *Properly executed, the drill is "shoot-pump-watch."* Stay on target a few seconds in case a follow-up shot is needed. Do NOT dismount the gun immediately. Do NOT pump it off your shoulder. Sometimes this is a hard thing to learn, especially for those with previous sporting experience. Repetition is the cure and you won't hurt a Model 870 by dry firing it. Run the gun assertively for proper function. A solid gun mount means smoother operation, which prevents short strokes. Maintain that head position and work toward smooth function. A properly drilled shooter has total command of the gun. The center of gravity remains forward and the head stays squarely in contact with the stock. The body language is one of dominance. A more timid approach will cause malfunctions, excessive recoil, poor control, and wild shots. Either *you* will run the gun or it will run you!

Now it's time to break out those dummy shells. Depending on how your gun is equipped, you may also need your shell carrying device. Establish a workable ammo-management strategy and use it during these exercises.

Combat load an empty gun (dummy shells). Here's a useful skill to perfect, especially with smaller-capacity magazines. From a home defense viewpoint it's also comforting to know you can quickly get your shotgun into action. Considering Murphy's Law, you'll need a designated shell source that is quickly accessible in low light. The classic pump gun technique involves grabbing a shell and rolling it under and up the receiver's ejection port side with the support hand. The idea has merit since the shooting hand can maintain a proper position. In our experience, it's a bit awkward and does have another pitfall. If the carrier (or shell lifter) of is fully elevated, the shell won't drop in the ejection port until it's manually depressed. We see this happen often. The ejection port is on the blind side for a right-handed shooter so it's easy to miss the problem, even in broad daylight.

The often taught under & up combat load technique.

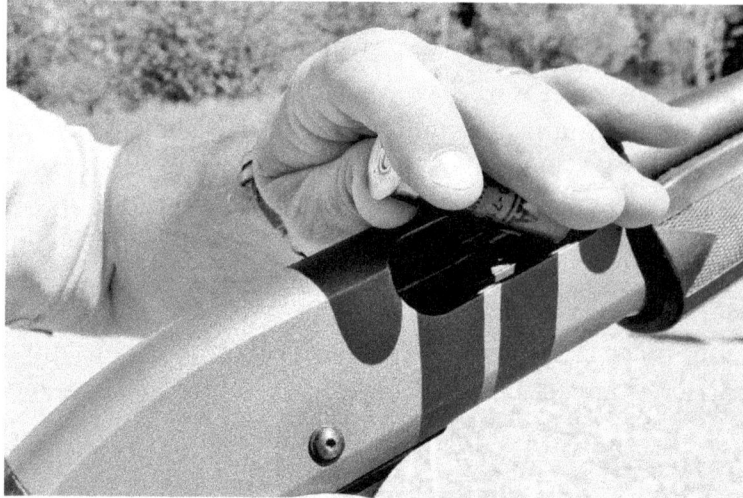

The over-the-top technique.

A fix involves canting the whole gun until the ejection port faces upwards. You can then bring your support hand over the receiver while watching the port. Better yet is an alternate technique - the one instance where we prefer strong-hand loading. The gun is empty anyway, so it's out of action. I like the small, two-shell plastic belt-mounted holders sold by Safariland (mine normally contain a pair of slugs). Locating one ahead of my right trousers pocket, it's easy to grab both shells at once. The first gets tossed in the side of the receiver and the second is stuffed in the magazine tube as the action is closed. Presto: two shells loaded in about two seconds. The technique can be modified for use with stock-mounted shell holders – handy during an emergency. If using a receiver side saddle, maintain a strong-hand grip and load with your support hand to maintain readiness. With either device you'll probably be better off grabbing just a single shell. The same applies for most belt-worn shell carriers, which should go on your non-gun side. By the way, you don't need to manually chamber a shell. Instead, just toss it in the receiver. The carrier will present it to the bolt face and it'll be shoved home when you run the slide forward.

Just toss a shell in the loading port. It'll chamber when the bolt goes forward.

Load to "carry condition" (dummy shells). Here's an agency term used by law enforcement officers to describe the general load mode of their shotguns. Some refer to it as "cruiser carry." In our case, we don't go on duty with chambered shells. The safety blocks the trigger, but an 870 has an internal hammer that is cocked upon cycling the action. If the gun is dropped or subjected to hard impact, the hammer can fall. A loaded chamber will result in a discharge if this happens. We therefore operate with full magazines, but unloaded chambers. Some civilians also store or carry shotguns in this mode. If unattended, such a gun should be fully secured from handling! "Carry" may involve bare hands, a sling, boat, vehicle, ATV, or aircraft. State laws vary so be sure to check first. In my home state, a civilian carry-condition shotgun is illegal aboard any conveyance.

To set up in carry condition, step "A" involves watching an *empty* chamber and running the slide forward until the breech bolt locks home. At this point, the internal hammer will be cocked and the slide will lock into battery. The safety should be engaged before shells are thumbed into the magazine. It only takes a second to trip the slide release, pump the gun, and chamber a shell. The distinctive sound gets everyone's attention. Some agencies modify the technique by pulling the trigger on an empty chamber. Once the hammer drops the gun can be pumped without using the slide release. It's simpler in theory under stress, but we don't like this version at all. Dropping a hammer anytime shells are present is just asking for trouble, and the slide may free float rearward.

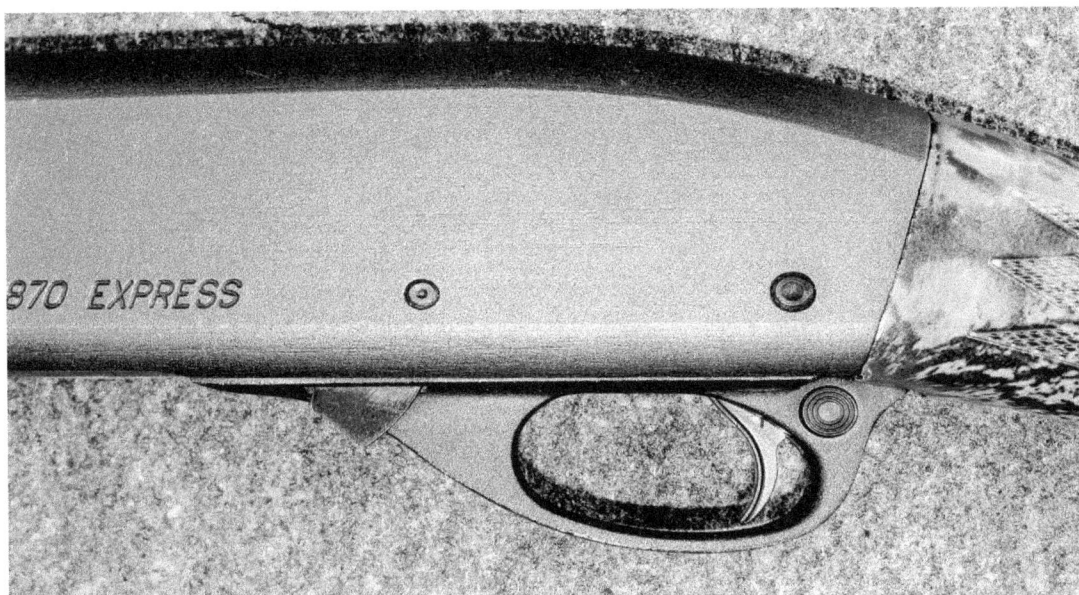

M-870 slide release (L) and safety button. Depress the slide release to unlock and open the action.

Verify carry condition (dummy shells). We do this often, including anytime the shotgun changes hands, or during vehicle placement. When the slide is cycled, the bolt runs over the hammer, cocking it, and engaging an "action bar lock." You can't crack the slide again until the trigger is pulled or the action lock is tripped. We call the tab-end a slide release because that's what it does. By depressing the slide release, the action can be carefully and partially opened. Open it only around an inch to check for an empty chamber. Go too far and a shell will spit out of the magazine. If that happens it's time to stop, roll the gun on its side, and let the shell fall into your hand. Practice makes perfect.

Transition to combat load (dummy shells). If trouble is brewing, one can unobtrusively position the trigger finger on the slide release. The forend can be cycled to chamber a shell upon depressing the release. There is no sound quite like it. If opportunity permits, a spare shell can be added to the magazine, thereby topping it off to full capacity. This subtle slide release trick is analogous to placing your hand on the grip of a holstered handgun, but much less obvious.

Down-load back to carry condition (dummy shells). Assuming the threat ceases, you'll need to clear the chamber again. A couple techniques are effective. In either case, the slide release is pushed to unlock the action. Moving the bolt rearward, the chambered shell is carefully extracted and caught. With the slide still rearward, any shell that popped onto the carrier is rolled out and also caught. In other words, two shells will need to be removed. If you're careful, you can cautiously bring the slide back just far enough to clear the chamber without tripping the magazine's shell stops. In that case, you'll only need to deal with one shell. Close the bolt on an empty chamber and top off the magazine. Verify correct carry condition one more time, just to be sure. The chamber should be empty and the safety should be "on"!

Completely unload without chambering shells (dummy shells). The technique is just an expanded version of the down load. After removing the first two shells, the gun stays open, the carrier is pushed up, and the magazine is emptied using the shell stops. This is a whole lot easier to do with the shorter, LE forend – which is why we like them. Without one, you can roll each shell out of the ejection port after bringing the slide fully rearward. At that point, a shell should pop out of the magazine and onto the carrier. Leave it in its down position, roll the gun on its side, and let the shell fall into your hand. If you first elevate the carrier, the shell will be trapped by a small ear designed to trap it during feeding.

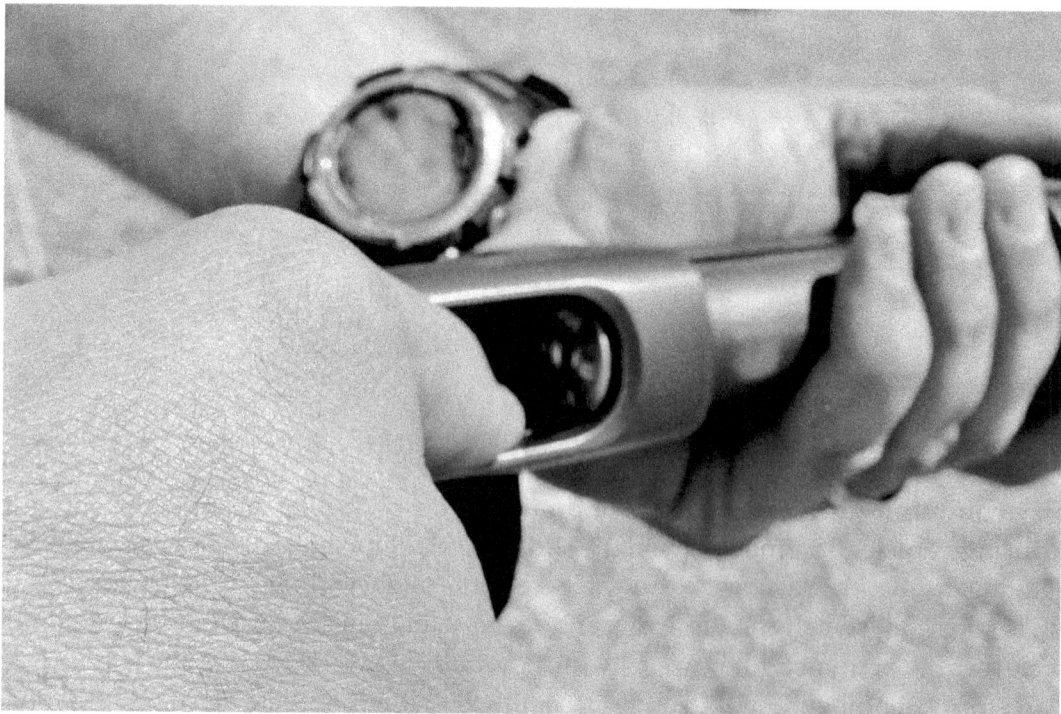

Depressing the shell stop will allow a shell to exit the magazine.

The process continues: the LE forend permits easy access and good visibility.

Important unloading caution: *It's easy to overlook a shell, especially with tubular magazines. Check it and check it again! We like the Scattergun Technologies lime green magazine follower for two reasons: It's visible and it has a nub you can feel in low light.*

LIVE-FIRE DRILLS

It's time to do some shooting. You'll need a safe, and ideally, private location. Thanks to regular gun mount practice, you should be able to handle a box or two of shells without suffering any recoil-induced effects. Lighter loads help and birdshot shells are cheaper. Reactive targets that fall or break won't know if they're being hit with garden variety promo loads or expensive 00 Buckshot. We employ 12-gauge 3 dram trap loads for basic shotgun training, firing them on reactive steel targets. We switch to "low recoil" 00 Buckshot shells for defensive purposes, which generate similar recoil. The less expensive birdshot shells permit more training and help minimize ricochet hazards.

Two loads with similar recoil. Birdshot costs less and still topples reactive targets.

Safety notes: *We don't shoot steel closer than 12 yards and we use don't use shot sizes higher than # 7 ½. Angles should be roughly 90 degrees, and commercial steel targets are recommended. They cost more but they're made out of harder alloys to resist cratering. Pock marks can shoot back, so damaged target surfaces are unsafe. Regardless, anyone in the area will need proper eye protection. When shooting slugs or buckshot, we stick to combat style paper targets.*

A Combat course employing stationary steel silhouettes. They get spray-painted between relays.

Knock down steel plates are lots of fun when combined with a shot timer. Stationary clay pigeons will work, but are fairly small for first time shooters. Keep ranges short initially, from 12 – 15 yards. By shooting with at least one other responsible party, someone can serve as "instructor" to help run the following stages. You can fire them with a bird-barrel gun, except for those involving slugs.

A more precise aiming system will be helpful when firing single projectiles, which is why the two-barrel option makes sense. In fact, if it's a smoothbore, the slug barrel will work fine for all of the following drills except aerial targets. Aiming causes misses, while pointing produces hits. A bird barrel is set up without rifle sights for this reason, promoting focus on the target. However, the rudimentary bead (sometimes two) generally lacks the precision needed to shoot consistent slug groups, especially at longer ranges.

If you have one, throw on the rifle-sighted, smoothbore slug barrel prior to shooting the first series of drills. For everything except slug shooting, employ "combat sighting", looking through the sights instead of staring at them. Fast hits will result from consistent head-to-stock placement.

Besides a gun, you'll want some means to carry extra shells. Since these drills are defensively oriented, pockets are your last option. A small 5 or 6-shell belt mounted bandoleer works well if positioned for access by the support hand. The capacity of this device and your magazine constitute a "basic load." Try completing all drills using just this supply, which will help you manage ammo. Grab a couple boxes of lower-cost promo birdshot shells like low brass #7 ½ loads. Put on your glasses and muffs, and standby…

Carry condition-to-fire drill. This drill is oriented toward those who keep a loaded shotgun for defensive purposes. We recommend an empty chamber, safety "on", and loaded magazine. The slide will need to be cycled prior to actual use, and the safety will need to be disengaged. Most hunters will carry their gun afield with a shell already chambered, but it's an invitation to disaster in very rough or slippery terrain, in vehicles, or with an unattended gun.

Others will maintain a completely unloaded gun (which we'll cover shortly). In any case, this first drill isn't a bad one to practice:

Load just one shell in the magazine and standby. On command or signal, hit the slide release, pump a shell into the chamber while mounting the gun, disengage the safety, and engage your target.

Repeat as necessary, firing at least 5 sequences (if not more). The goal is familiarity with the steps needed to efficiently get your gun into action. Loading only one shell saves time, and will provide a safer opportunity to identify any control issues. Start at five seconds but work toward three.

If things are going smoothly, load two shells. You'll only need to activate the slide release for the first shot, pumping the second shell in immediately afterward. It can be fired on a separate command, or immediately.

Sustained-fire drill. Now it's time to fill the magazine. This drill is an extension of the previous one. We'll still begin in carry condition, but activation of the slide release will only be necessary for the initial shot. Let's first consider a few key principles that can be applied to this drill:

➤ **Shoot and replace:** We put our shooters up against an array of multiple steel silhouettes, verbally designating targets. Shooters top off their magazines as opportunity permits. For example, if two shots are fired, we'll be looking for two more shells to enter the magazine, striking a balance between maximum capacity and threat coverage. Reloading is subservient to readiness. Well-drilled shooters can

Sustain magazine capacity using the support hand. Shoot & replace.

effectively negate limited magazine capacity and the gun is never out of action until all shells are expended.

- ➢ **Load with your support hand:** The shoot and replace technique works well when the support hand is feeding shells to the magazine. Using this method, the gun can be topped off without ever leaving the shoulder. Our pet peeve is the upside down gun, strong hand, duck hunting reload. It's okay in a blind but it will take you out of a fight during deep doo-doo.

- ➢ **Ammo management:** Properly located shells will permit support hand reloading. Several means are available, including bandoleers, two shell belt holders, multi-shell dispensers, and gun mounted systems. As a righty, I like spare shells on my left side, carried in a belt-mounted system. Even though we use side saddle, gun mounted shell holders on duty, it's not my preferred choice. I'd still pick one over a stock-mounted elastic shell carrier, though. Dumping shells in a support side pocket is quick, but our least favorite option. They may be inaccessible from some positions, and we've seen a number of foreign objects go in magazines by mistake. These include lifesavers, anti-acid rolls, and cigarette lighters. Last observation: It's an all too easy way to load shells backwards. Good luck if that happens! A carrier that properly orients shells is the best way to prevent the problem.

It's time to make some noise! Load to carry condition and standby with the safety on. A smart camper would sneak the trigger finger forward to the slide release, in preparation for the first command. The slide can be cycled en route to the shoulder, and the safety can be disengaged last.

Multiple targets will make things more interesting, especially if shot in a random but designated sequence. Begin with just singly called targets, advancing to designated double or triple engagements. In between, return to a "ready" while covering the array. Your trigger finger must be outside the trigger guard on the frame. Don't let it curl behind the trigger guard.

Meanwhile, reload the magazine as opportunity presents, not forgetting the "shoot-pump-watch" sequence. It's possible to top off your magazine while maintaining a gun mount, but it's easier from a ready position.

Thought: If you began in carry condition, the chamber was empty. Upon chambering the first shell, the magazine lost its total capacity. Even if you only fired one initial shot, you could reload two shells!

Always pump the gun on your shoulder, and don't immediately dismount it after a shot. Instead, be prepared for

Pump the gun on your shoulder and don't immediately dismount. You may need a fast follow-up shot!

a possible follow-up shot with your "instructor" playing devil's advocate. As the shooter, you'll need to load when opportunity permits to stay in the game. Run through a reasonable "basic load" consisting of a real life shell quantity. In our agency's case, this will be around 13 shells. Develop a system and progress through your spares while maintaining awareness. As you sweep your supply with the non-shooting hand, you can keep track of remaining loads.

"Did I git 'em?" We hope so because you're out of the fight!

Assuming you've been able to consistently shoot and reload, at some point your external shell supply will become exhausted. You can then try counting down from the magazine's capacity. The goal is to engage each designated target, without encountering the loudest defensive sound in the world; "CLICK."

Target engagement drill. This variation of the previous drill is designed to promote fast shooting. One difference is that it begins with a fully loaded gun. Depending on your anticipated load preference, you can start in carry condition, or just throw a shell in the chamber and fully load the magazine. Either way, start with the safety on.

Plot twist: Carry condition shooters will now encounter a subtle variation. After chambering their first shell, the magazine will be down one. Why not top it off prior to beginning?

From a ready position, think about what you'll need to do for a fast response. With a chambered shell and the gun on "safe", mount and fire on the start signal. Strive for a hit within 2 seconds, engaging each target from a separate mount. With more practice you should be able to topple 12 yard, 8" plates in 1 ½ seconds. Reload between shots until your basic load is expended. Try to count down so you know when you're out.

Emergency combat load drill. Ideally, we should be able to sustain our magazine's capacity, but either no opportunity will present itself, or human nature might prevail. In other words, you may wind up shooting the gun dry even though spare shells are still on hand. Many folks will also keep a shotgun for home defense, but store it unloaded (including me). For this drill, start with an empty gun and your action open. Upon signal, toss a shell in, run the slide home, and make a hit within 3 seconds. Next, advance to a pair of shells and two targets within 5 seconds. Quickly chamber the first and stuff the second in the magazine before engaging your targets.

Try to remember to leave your slide rearward between each series. It's harder than you think, and there's a good chance you'll run it forward due to programming. Ideally, you'd know you were out with the presence of mind to NOT pump the gun. That's easier said than done, so you can also begin a few series with an empty gun, still in battery. Mount it, dry fire, load, and get back in the game. Depending upon your home storage plan, either version (or both) should have value. The drill can begin with the safety either "off" or "on." In the middle of a fight, it would probably remain off.

Combat load and fire course. A progression from the previous drill, this one involves 5 shells and as many targets. Start with an empty gun on "safe" and its action open. On command or signal, load all 5 shells before engaging each target. Begin with a goal of 20 seconds and strive for 15. You'll be busy!

Combat load & fire course with knock-down plates. Each must fall to count as "hit". Note the pellet marks.

This drill isn't the best tactical option, but it does promote good gun handling. If you're weak on loading skills you'll fumble shells. Without a solid gun mount, your head and the stock will part, resulting in wild shots. With *lots* of practice, you may be able to crack 10 seconds. We use three knock down poppers and two 12" disks. An electronic shot timer is ideal and will record any glitches. As a right-handed shooter, I shoot the array from left to right, using recoil to assist in target transition. In the real world, the most immediate threat would be priority number one.

Man-on-man (or person-on-person). Use the same 5 target array and line up two shooters. Each begins with an empty, on-safe gun and 5 shells. On command, each shooter must load and engage two assigned targets. Downing of a third "stop" plate will determine the winner, but it can only be engaged *after* a shooter's assigned pair are down. Generally, the best strategy is to load 3 shells and get cracking while being prepared to load extras. In our experience, all bets are off on this drill, since everyone is susceptible to Murphy's Law. It's a great spectator drill and an audience is the best way to referee the outcome.

Tip: Work on controlled and efficient gun handling. Going flat-out won't work without lots of practice. Instead, think about exactly what steps you'll need to exercise first, and then smoothly execute each one. *Smooth is fast!*

SLUG SHOOTING

It's time to slow down and re-group. You won't achieve optimal results without good sights that are properly zeroed for both you and your chosen slug load. Individuals often see their sights a bit differently, requiring personal adjustments. Different loads may also deviate from an established zero.

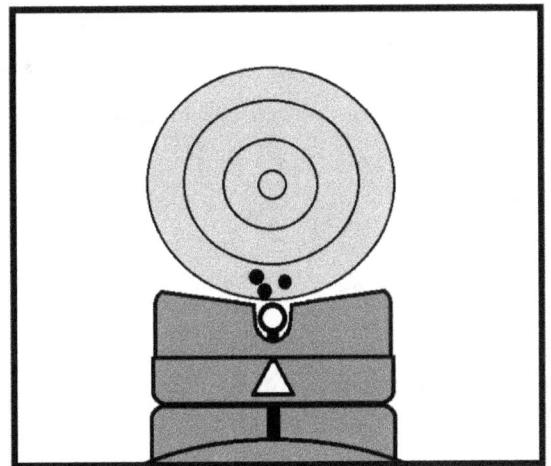

Sight alignment and zero at 50 yards. The 6:00 hold is a bit more precise. At 25 yards slugs should strike higher or closer to center.

Sighting in. Remington's slug barrel open sights will get you by in daylight conditions. We zero them at 50 yards, remembering to move the rear sight in the same direction the slug must go. In other words, if your groups strike low, the rear sight will need to be raised. Be sure to bring the right tools with you, which may include an Allen wrench or small screwdriver set. Most Remington rear sights use an inclined base marked with reference lines. The moveable sight body has a corresponding witness mark, and slides up or down the ramp after loosening a set screw. Note its position before making any changes. A separate screw on top permits windage adjustments. Move it in the direction you want your groups to go.

If your slug barrel has a fixed choke, it's probably either cylinder bore or improved cylinder. If it is equipped with choke tubes, install one of these constrictions, or better yet, screw in a rifled choke extension. You'll need some sort of rest to minimize human error. A hard surface can give false reads, but a typical shooting bench setup may beat you to death. From a seated position, the gun may greatly exaggerate recoil. Since we often deal with a quantity of guns, we stack a couple milk crates on top of a shooting bench to serve as a support hand stabilizer. We then carefully shoot from standing, letting recoil transfer our balance from the front foot to the rear. It's not the best way to determine accuracy, but does replicate offhand shots with greater shooter comfort.

When comparing the relative accuracy of different loads, a better rest is necessary. In that case, use a chair, shooting bench, and rests with soft surfaces. Try to adjust everything for straight-up shooting to help absorb recoil. You might want to drape a folded towel over a tripod rest to simulate hand tension. It will also prevent damage to the gun caused by impact with the rest in recoil.

Sighting in off a shooting bench during real-life weather conditions, prior to a January deer hunt.

The best weather conditions to achieve a repeatable zero are chilly with an overcast sky and calm wind. Cooler temperatures ensure less velocity deviations from extremes. Cloudy sky provides uniform light and still air improves groups – especially at longer ranges. Using iron sights, you'll often shoot "into the sun", meaning if it is shining from your left shoulder, groups will likely strike a bit left. So-called "neutral" conditions help mitigate these effects. You'll need to work on trigger squeeze, letting each shot break as a surprise. Let the gun recoil and follow through, focusing on your front sight. Fire at least three careful shots before making any sight adjustments.

Once your slug groups are striking on top of the front sight at 50 yards, note the location of the witness marks on the rear sight's elevation and windage scales. You might want to span them with pencil marks to see if the parts move. Recoil can cause the sight body to creep up the base ramp over time, causing shots to strike high. A bit of blue threadlocker or nail polish is good insurance once your final adjustments are made. Just be aware that a change in loads will probably require different adjustments.

Drills and trajectory. After your slug zero is established, you can shorten the range and shoot a few at 25 yards. The previous man-on-man drill is a great motivator, requiring solid marksmanship skills in order to prevent flinching with hard-kicking slugs. Balloons make great targets. So do milk jugs, big cans, static claybirds, or paper silhouettes. To cure a case of flinch-itis, have a partner load your gun, mixing in a few dummy shells. It's quite revealing and, in our experience, also common to flinch on a dummy shell. Lots of off-range dry firing helps and it won't hurt a Model 870.

A 50-yard M-870 smooth-bore slug group. It's a bit high but still adequate.

RIFLED-SLUG GROUPS FROM A CYLINDER-BORE 12 GAUGE

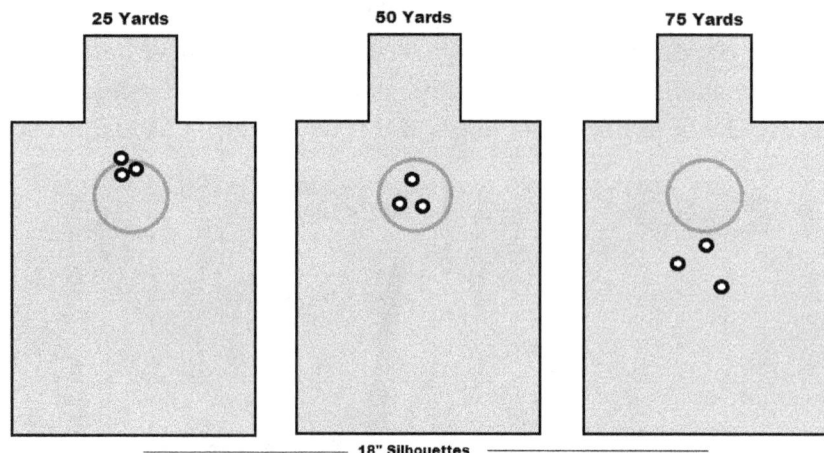

| 25 Yards | 50 Yards | 75 Yards |

———————————— 18" Silhouettes ————————————

A rough depiction of 12 Ga. 1 oz. slug impacts with a 50-yard zero.

Eventually, you can move back to 50 yards again, and then stretch the distance to 75 yards, or beyond. It's worth firing a few careful groups back to 100 yards, noting your drop and accuracy. Wait for calm conditions when pushing the envelope. Those shooters with special, rifled barrels and premium sabot slugs will be in league of their own. In such cases, a 100 yard zero may be feasible. One thing to remember: Rifled barrels will totally blow shot patterns.

BUCKSHOT

Having nailed a useful slug zero, it's a good idea to fire some buckshot patterns. We're looking for two things: impact points and pattern densities. You may see some discrepancy between slug and buckshot strike points. If so, it's worth noting.

Time spent at a pattern board can also be educational. Don't expect to gain a handle on performance from only one or two patterns. Sometimes buckshot is a crap shoot, with widely varying results from one shot to the next. We recently shot a few stunning patterns from fixed I/C Remington M 870s equipped with 14-inch barrels and Federal Tactical 00 buckshot. From around 20 yards, all 9 pellets formed tight 8–10" clusters. Here's the kicker: Since we only fired a few shells, it could just be an anomaly. On the other hand, after firing LOTS of patterns during controlled conditions with more conventional barrels, a few recommendations seem safe.

Chokes and patterns. Try a modified choke if firing 00 Buckshot. Some of the newer "low recoil" loads use modern wads, and will even do well through a more open improved cylinder constriction. Five shots are a minimum to assess performance, but I'd feel a lot more confident after firing at least 10 patterns from a given distance. Ideally, fire each series in 10-yard increments, out to 50 yards. You'll probably see a few random examples, but you should also be able to gauge overall performance. More than likely, 40 yards will be your useful outer limit, although some of the most expensive, premium loads may do better. In any case, the pellets rapidly shed velocity and energy, which limit their terminal performance, but at close range, you'll have serious punch. Shoot a few patterns inside 10 yards and you'll see that buckshot can be fired discriminately - *if you know your gun and patterns.*

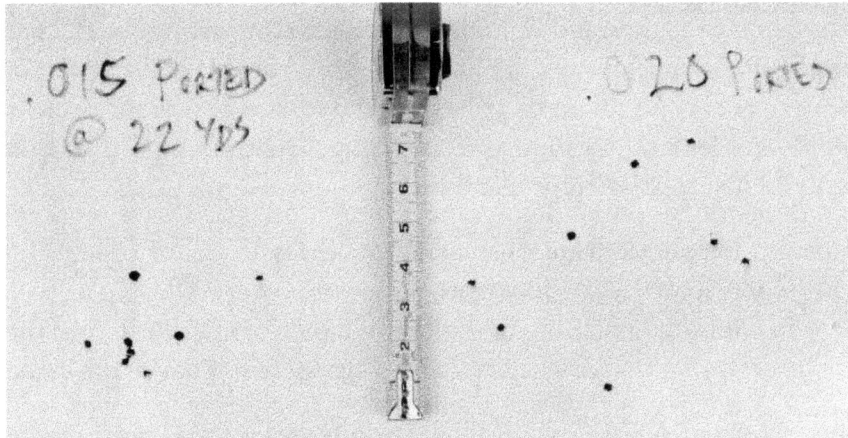

Tight patterns fired through Angle-Port chokes with 8-pellet tactical 00 Buck. The left one was through a tight I/C and the right was through a Modified constriction.

00 BUCK PATTERNS FROM A CYLINDER-BORE 12 GAUGE

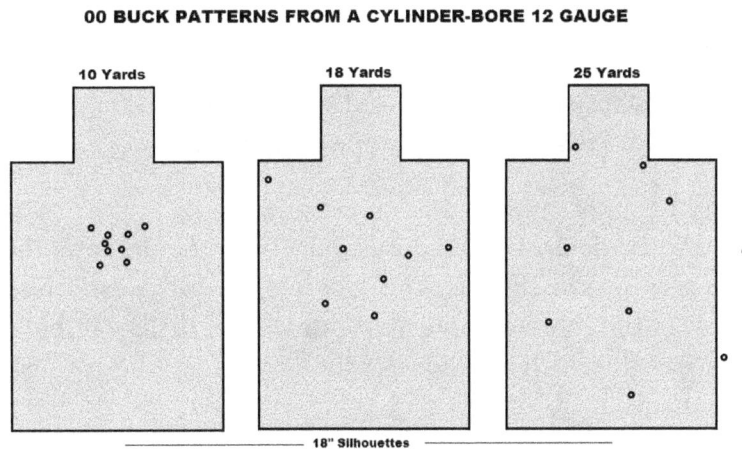

Typical 00 Buckshot spread from a cylinder-bore barrel. Patterns will expand roughly one inch per yard of travel.

Ammo changes. Here's a scenario to consider. You're toting a pump gun full of 00 Buckshot, but also have a few slugs. A sudden but distant threat requires an immediate response. What do you do? We've messed around with this situation, and here's our K.I.S.S.-based option. Inside of 60 yards, you might as well launch a fusillade of buckshot, preferably while moving towards cover. There's a good chance that *some* of those pellets will connect. At that point, you'll be a few shells down, leaving room in the magazine for slugs. Stuff some in and resume firing. Your next shot will be another buckshot shell, but the slugs will quickly follow. It's not the most Ninja-like response, but it's fast and simple.

Buckshot patterns become excessively thin beyond 60 yards. Luck may prevail, but an immediate change to slugs is a safer bet. In that case, several techniques exist. The simplest is to eject any live shells and replace them with slugs. Using a Model 870, bringing the slide fully rearward will extract a chambered shell, but cause another to feed into the shell carrier (or lifter). You can shake both out and load two slugs. Toss the first one in the ejection port and run the slide home. If time permits, thumb a second slug into the magazine. Location of your spare slugs makes or breaks this operation, and is why I carry two in a belt-mounted holder accessible to my strong hand. I can grab both in one smooth motion quicker than you can say "big holes." A well-trained shooter can block the magazine while pumping the gun. A slug can then be tossed directly into the action. Some agencies load their magazines one shell down, leaving room to stuff in a slug. The trick would be to remember this at a moment of truth, and I'd just as soon have a full tube.

Regardless of the chosen technique, true proficiency requires LOTS of practice! You'll also need a practical means to segregate loads, accessible from different positions. Using a receiver-mounted side saddle shell holder, you could reverse the orientation of a pair of slugs and find them in the dark. I'd put them in the most rearward slots. A stock-mounted sleeve could serve the same purpose.

As mentioned previously, we're not fans of alternating buckshot and slug magazine loads. It's an idea that sounds good in principle but can fail in practice. A well-drilled operator will shoot and replace, but that can complicate matters.

TRAINING ON OTHER GUNS

The drills we covered are geared toward magazine-fed systems, and many can be tailored to semi-automatic designs. However, we did discuss a few simple designs like single shots and double barreled guns.

I'd skip single-shot shotguns for defensive purposes, but for home defense, the lower-cost side-by-sides may have value. They're extremely simple to operate, and it doesn't take long to load a pair of shells. In that case, most of the more complicated pump gun drills can be skipped. Instead, just cherry-pick those with direct value such as ready positions, gun mounts, or fast load and fire stages.

It takes lots of practice to execute fast break-barrel reloads, but it is possible. Believe it or not, a great source for such skills is in "Cowboy Action Shooting." You can see the top shooters in action by watching the outdoor channels. You'll gain a whole new appreciation of the term "fast"! Switching loads is pretty simple, too. The biggest drawback is their limited capacity.

Mossberg's new "Thunder Ranch" over & under 12 Ga. is geared towards defense, with Picatinny rails and a modest retail cost of $565.

AERIAL TARGETS

We often hear complaints about difficulty hitting clay pigeons. Well-schooled rifle shooters are typically among this group. The problem often boils down to aiming. It just doesn't work on aerial targets! Instead, we need to point a shotgun, which requires a real leap of faith. Unlike precision rifle shooting with its various mechanical aspects, wing shooting success is based on movement and form. A plain, bead-sighted bird barrel also helps. The addition of sights only increases the urge to aim. The result will be hesitation, over-thinking of leads, and lots of missed targets. "Lead" is the distance held ahead of a moving target, in order to make a hit. In other words, we'll need to shoot at a spot where the target will be when the shot pattern arrives.

There are several popular formal clay target games. Each is governed by established rules and managed by a national organization. Within each game one may encounter a couple versions. For example, "American" Skeet is shot with the gun pre-mounted, whereas the "International" game starts from a ready position. The standard target launch command is "Pull!" and targets must be chipped or broken to count as hit. The best shooters in each discipline are very good, running near 100% hit averages. Shoot-offs are common during champion events. Guns are highly customized and the competitors are very serious. Most of the guns will be either pricy target-oriented over and unders or autoloaders. However, it's still possible to have a good time with the gun you have.

TARGET LOAD

AVERAGE PELLET COUNT

SHOT SIZE	⅞ OUNCE SHOT	1 OUNCE SHOT	1⅛ OUNCE SHOT
7½	306	350	394
8	359	410	461
9	512	585	658

RECOMMENDED APPLICATION

DISCIPLINE	SHOT SIZE	CHOKE
TRAP	7½, 8, 8½	MODIFIED, FULL
SKEET	8, 8½, 9	SKEET, IMP. CYLINDER
SPORTING CLAYS	7½, 8, 8½, 9	IMP. CYLINDER, MODIFIED
PRACTICE	7½, 8, 8½, 9	IMP. CYLINDER, MODIFIED, FULL

A handy clay-course reference chart to help select target shells and chokes.

Trap. This game is a fairly long-range event, fired at random, outgoing targets. A series of 25 standard-size 108mm-diameter "clay pigeons" are thrown in a fan-shaped but outgoing series, from one protected downrange bunker. Five shooting stations are arranged 16 yards behind the oscillating thrower (the trap), in a semicircular layout. A squad of five shooters steps up to their respective stations, firing 5 shells from each spot. They eventually rotate through all five stations until each shooter has fired 25 shells. Handicap stations are also located further back to increase the range. Compared to other games, Trap is fired with heavier loads and tighter chokes.

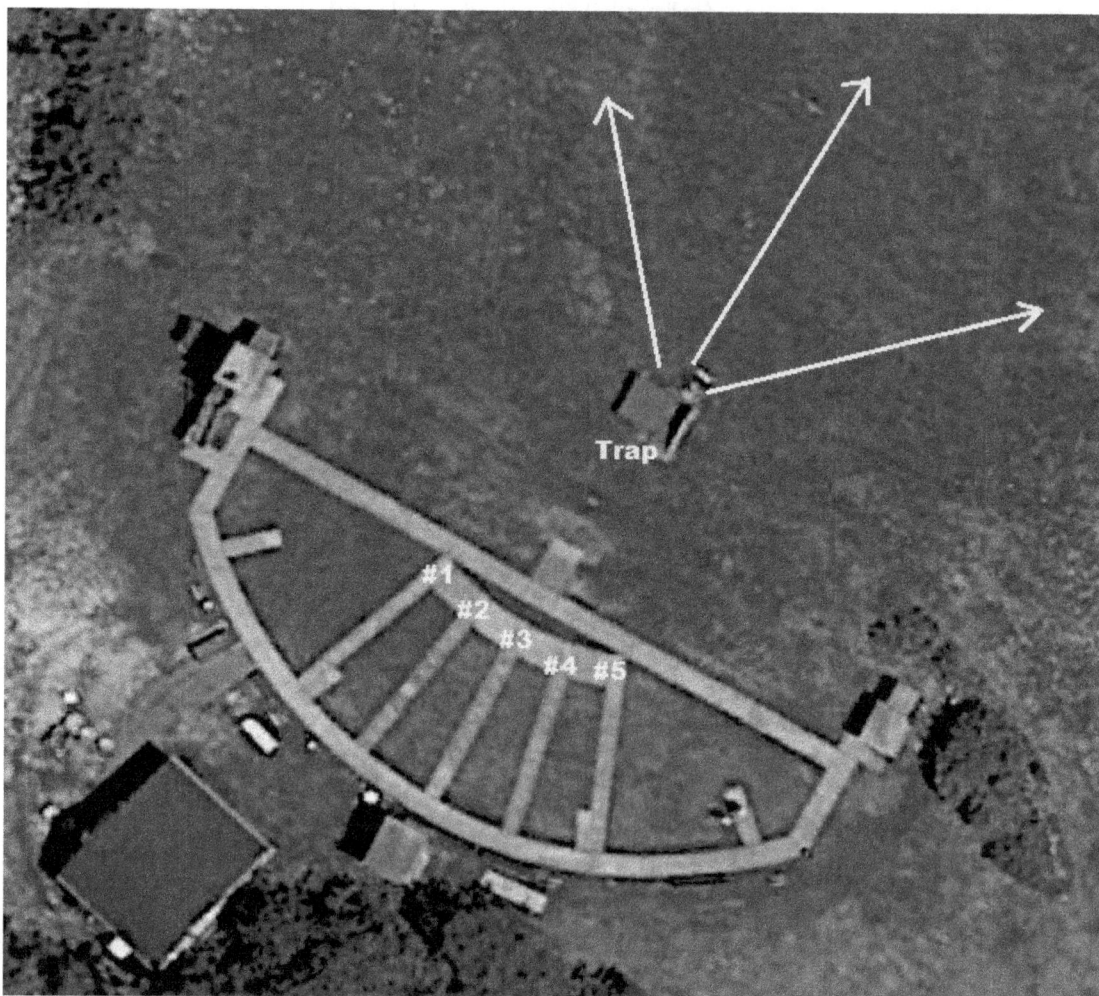

Overhead view of a regulation trap course, showing target paths and shooting stations.

Skeet. Originating as an upland bird hunting simulation, this game involves closer targets shot at from various angles. Once again, a series of 25 standard targets are thrown, but unlike Trap, they emerge from two locations on pre-set paths. Opposing high and low "houses" throw single targets, or simultaneous pairs. A squad of shooters fire from eight total stations which are arranged in a semicircular array. Each shooter begins at "Station One" and the squad moves to the next station after all shooters have had their turn. Initial targets are outgoing or incoming with shallow angles; however, as the stations progress, angles increase up to 90 degrees, and then decrease again. The last targets

(Station 8 and 9) are engaged at very close range between both houses as fast incomers. Since most targets are shot inside 21 yards, more lively guns with open chokes are preferred, along with lighter loads and shot as small as #9s. Regulation Skeet is also a four-gauge event with an interesting .410 class.

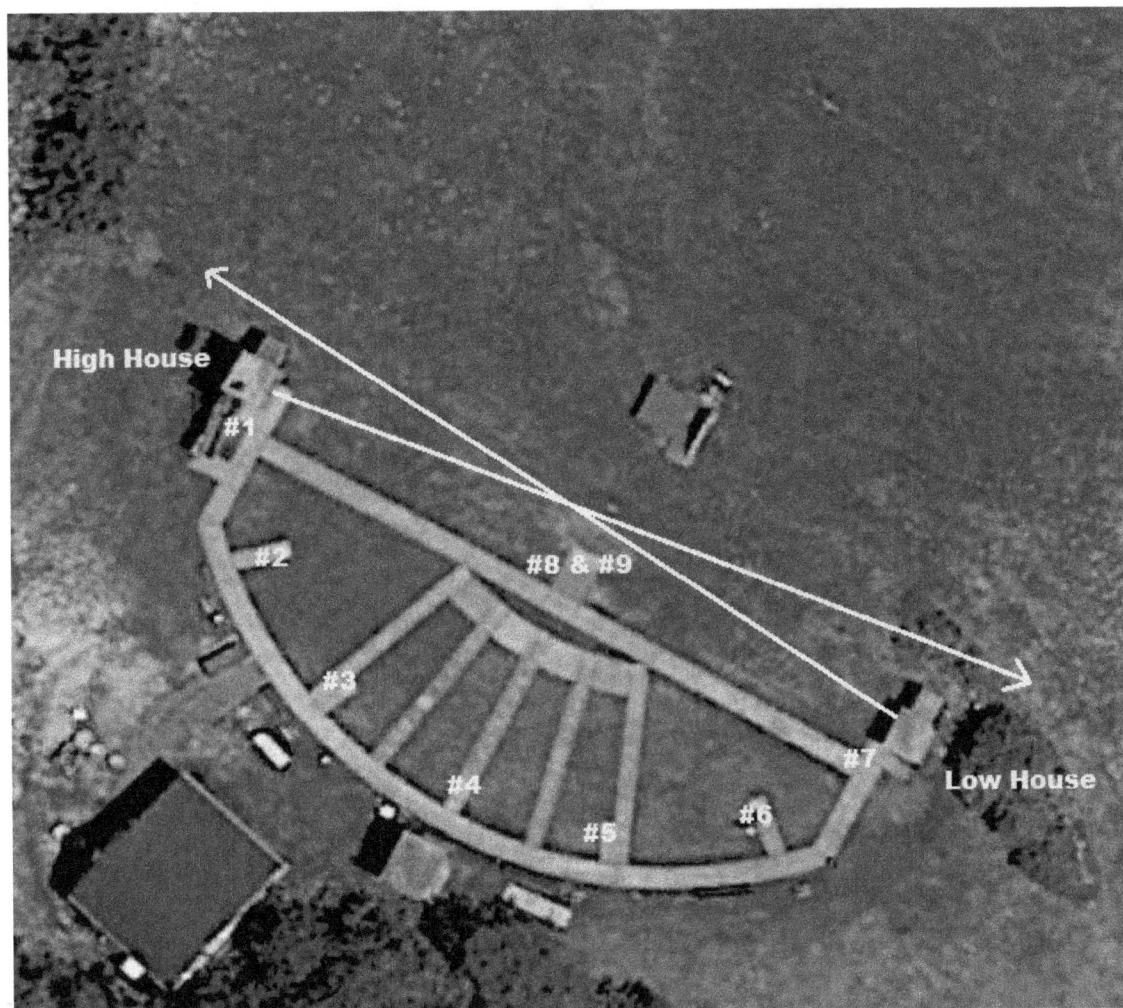

The same field, adapted to skeet. Five-stand might employ both courses.

Five Stand. By combining some aspects of Trap and Skeet, shooters get a chance to fire at an interesting spectrum of targets. Like Trap, there are five shooting stations arranged in a linear fashion. However, like skeet, much sharper angles and multiple targets are encountered. Each station has a posted target sequence which varies from one to the next. Shooters rotate through each station until their round of 25 targets is completed. Although 2 shots are permitted at each target, shooters are limited to only 2 shells. In other words, this strategy won't save the day on double targets. These are also interesting, consisting of small, fast 60 mm "Minis", mid-size 90 mms, standard 108 mms, thin, flat "Battues", and heavy-duty "Rabbits" that bound edgewise across the ground. Shooters may use a bit more choke and shells appropriate for Trap or Skeet, depending on each station. It's a lively event, with highly challenging variations such as the European FITASC.

An assortment of claybirds including 108mm Standard, 90mm Midi, 60mm Mini, Battue & Rabbit targets. Note the pellet holes in the Rabbit.

A Rabbit is thrown on edge, bounding across the ground. It needs to be tough enough to survive repeat bounces. This target survived multiple #8 pellets. For reliable breaks, switch to warmer #7 ½ trap loads.

Sporting Clays. Picture golf in the woods with a shotgun combined with the targets from Five Stand. A squad of shooters travels through a course consisting of many stations, often spread out in a wooded setting. This game was developed to simulate hunting, so there are no standard target presentations. Instead, each course designer is free to design interesting and unique targets using the full gamut of different claybirds. Shooters can expect outgoing, incoming, crossing, rising, or falling targets as singles or pairs, often with trees or other obstacles to overcome. Each station has a posted series of targets and the first shooter up can call for an initial non-firing sample. A round consists of 50 targets and, like Five Stand, the same 2-shot rules apply. Some shooters obsess over chokes and loads while others just use #8s and a hunting gun equipped with improved cylinder.

A fast left-to-right crossing target (inside circle). There's a lot of air around 'em!

Simple hand-throwers provide lots of entertainment. The old wooden-handled Remington hand-trap has been in service for decades.

No moving parts other than a strong arm!

Informal shoots. Anyone fully into the above games will amass a large and expensive assortment of gear, not to mention dedicated guns. However, there really is no reason why plenty of fun (and good experience) can't be enjoyed with friends in a field or gravel pit. Several reasonably priced manual traps are available permitting an informal layout. Some are staked to the ground while others are mounted to tires, plugged into a trailer hitch, or designed with a built-in seat. We've had good luck bolting a basic version to a homemade sawhorse-type bench, set up low enough to reach a claybird case. Sometimes we'll round up extra traps and take turns as throwers or shooters. For just a few bucks you can buy a simple, handheld thrower and introduce some unpredictable birds. Some folks may decide to pool their resources and buy an electric trap. Some are reasonably priced. We only have experience with two more costly magazine-fed "Auto Chuckers", in use for 10+ years. Each costs around $1,800, runs on a deep cycle 12-volt battery, and is cycled by a radio controlled remote. Battery life is great, and we can run both machines off one remote either separately, or while throwing true pairs. One big advantage is the ability to safely locate a machine down range for incoming targets. We'll also combine these machines with other, simple throwers, thereby creating our own Five Stand "Redneck Clays" course. After firing in the same area over a period of time, a large debris field of broken clay targets will accumulate – something to keep in mind. Fortunately, you can buy biodegradable claybirds!

A basic Trius Trap, bolted to a home-made seat. It works well as long as the operator stays clear of the energetic throwing arm.

A pair of radio-remote Auto-Chuckers. The controller will run both machines, which feed from 65-bird magazines. True double or incoming targets are possible – for a price.

The radio-remote controller, which runs on a 9-volt battery.

WING SHOOTING SKILLS

It's worth spending some money for a professional lesson, even if you already have experience. Just a few hours spent on a skeet field with a pro will prove extremely valuable. You'll learn about pick up points, hold points, and break points. A knowledgeable coach can help you understand the best place to visually lock on a claybird, the optimum spot to point your muzzle, and the area in which to break the bird. Your foot and body position will also play key roles. Employment of these skills will result in smooth and controlled gun handling

What about judging lead? We can't shoot directly at a 25-yard crossing target and expect to hit it. We'll need some forward allowance, just as we would when throwing a football to a broadside running receiver. Here's the funny thing: You'll probably be able to make an accurate toss without any conscious calculation at all, but when it comes to clay targets, many of us will obsess over the *exact* lead. As a result, we'll over-analyze the shot, hesitate, and miss. Like that football pass, it's more a case of letting the force be with you. Believe it or not, what matters more may be how you develop your gun mount and swing than your precise degree of lead! A good many shooters can hit flying targets with no conscious idea of their specific technique. On the other hand, the most disciplined claybird competitors often intentionally employ several techniques, matching one to a specific target. The most common techniques follow.

Swing-through. Using that 25-yard crossing claybird as an example, let's assume it's traveling left-to-right at a 90-degree angle to the shooter. Total focus should remain on the target while the muzzle acquires the target's path. The gun starts from behind and accelerates through the bird. As a bit of space develops the gun is fired, developing a built-in lead. The exact amount will depend on several factors including the target's speed and the gun's inertia. It's not an exact science so much as a leap of faith. Swing-through may be the most common technique, whether consciously employed or not. Plenty of casual shooters and hunters use it without any idea at all. The pros often use it on quartering or very close targets. Since there are several variations, this is a simplified rendition.

Pull-away. Instead of coming through the target from behind, the muzzle locks on and tracks it until just before the shot. At that point the gun is accelerated and fired, which automatically introduces lead. Again, the exact amount of developing space depends on a few variables. Pull-away is sometimes used on more difficult, long-range, crossing targets to improve swing control.

Sustained lead. Using this technique, the shooter's muzzle is inserted ahead of the target, maintaining a constant gap. Exactly how much is once again a variable. A very rough rule of thumb is two finger's worth. On an average-speed crossing claybird at 40 yards, extending two fingers ahead of the target should translate into a gap of roughly 7 feet. The operative word is "roughly." It's not an exact science and any attempt to make it one will probably result in aiming (instead of pointing), followed by a stopped gun, and a miss behind the bird.

Other factors. Some shooters are wired for one technique more than another. Still, for anyone bent on true mastery of wing shooting, it's worth understanding them all. And foot work plays a huge part in the initial setup for a shot. It's all too easy to run out of swing by improper body positioning.

Wide foot placement doesn't help, nor does a sharp shoulder angle relative to a target's path. For a right-handed shooter, facing too far to the right in a rifle-like stance may easily result in a checked swing on a right-to-left crosser. When in doubt, try naturally facing your belt buckle toward your intended break point. Maintain your footing and assume a ready position by only moving your upper body. This may seem strange initially, but will feel a whole lot more natural as your body straightens out to make the shot. Think "move-mount-shoot."

The wing shooting information described above is really just an over-simplified primer. Again, lessons from a real pro are a *much* better bet. Those less informed will often provide "helpful" coaching which is not always accurate. Everyone worries about their correct lead and your sandlot coach may offer an opinion. But, what if you also have a built-in flinch - something that is very common? In that case, your forward allowance may not really be the problem at all. Instead, targets will be under-shot. Lifting one's head off the stock has the opposite effect. Sometimes one error will cancel the other one out, but that's plain luck (and unadvisable). Eye dominance issues can cause major pointing errors and some often repeated diagnosis is actually incorrect. With the price of shells and claybirds being what it is, professional instruction becomes a worthwhile investment in time and money.

Once you have a handle on your wing shooting basics you'll gain new confidence. The ability to connect with fast-moving targets is the essence of shotgun shooting. That's why, during our multi-day, combat shotgun program we include this discipline as a finishing drill. One big difference: They don't get to call "pull." Instead, we run a two-shooter team against an un-called flurry of claybirds that emerge from two machines. The shooters need to stay in the game by reloading and communicating. They're pretty busy, too!

With a handle on the basics, you can polish your skills and have a whole lot of fun with nothing more than a manual trap and a couple cases of clay birds. At that point, you'll have become a good, all-around shotgunner.

ADVANCED TRAINING

The previous drills should turn you into a fairly competent gun handler, but professional coaching was recommended for wing shooting skills, and the same logic applies to more serious endeavors.

Defensive competency involves other factors such as tactics, and just being "a good shot" won't necessarily save the day. Effective movement, use of cover, and room clearing strategies are but a few tactical examples which may need to be addressed in adverse environments. Beyond tactics, lights, slings, and other gear are always evolving. We can't possibly cover everything in one manual and even if we could, safe integration of these aspects is best developed with knowledgeable supervision. In other words, a good hands-on school is the way to go.

Shooting on the move or from cover introduces new challenges, even in bright light, with two working hands. Add fast sling dismounts, doorways or vehicles, and the tempo will increase. Remember my crow hunting story about tipping over backward while seated? We've seen macho shooters lose their balance while kneeling from low cover due to improper techniques. It's pretty scary on a range and not conducive to longevity in serious social encounters. *Everything* changes in prone and some interesting stoppages may develop as well. Fast vehicular exits introduce muzzle management concerns and other problems. Turn out the lights and you may encounter shells loaded backwards into magazines. Managing two or more different loads adds further challenges. The list goes on.

Many of these issues could be encountered in just one event, so the extra insurance offered by professional training is worthwhile. There are plenty of good schools throughout the country – certainly more than I'm aware of. A good starting point just might be the well-known Gunsite Academy located in Arizona. Tuition is a bit north of a grand, but covers three well-spent days.

You'll probably get what you pay for.

CHAPTER 11

SECURITY, CARRY MODES, AND CUSTODY

Survival Guns: A Beginner's Guide contains an entire chapter dedicated to secure storage methods. A number of options are covered from gun safes to smaller lock boxes. One problem with shoulder-fired guns is that they just won't fit in most of the popular secure but rapid-access containers, which are usually geared toward handguns. Another concern is that unlike handguns, they won't fit in holsters. Since carry normally involves at least one hand, it can be limiting. Any separation from the user raises concerns related to security. In other words, we have a few final things to think about.

An assessment of your lifestyle is a good place to start. Unauthorized access is a recipe for disaster, starting with theft and going downhill from there. Loss of a child to an unattended gun is the ultimate nightmare, and one you'll be accountable for. Even if your kids are gun savvy, their friends may not be. Without children in the equation, unless a firearm is in your direct custody, some measure of security is still recommended.

SECURITY

Out of every means available, a gun safe tops the list. For home use it's hard to beat one. Any firearms going in it will be safer if unloaded. The training section should convince you that it doesn't take long to get a shotgun into action. Shells should be stored separately so all of your security eggs aren't in one basket. One problem is quick access. Spinning a combination lock in the dark while under duress will probably not work. Pushbutton entry locks may be better with regular practice. Depending on personal circumstances, you could open your safe when home and lock it when you're not. Kids remain a concern, as does your memory!

Secure but not very portable!

Also, bugging out with a 650-pound safe is just not feasible. It will, of course, lend some measure of protection for anything left behind.

Single gun owners have less baggage and might consider a heavy-duty but portable locking case. Unlike a full-size gun safe, its portability means it could be lugged off by others, so one new compromise is a series of smaller hybrid safe/fast-access gun lockers. They're really just larger versions of handgun locking boxes, although some are heavy-duty. They can be bolted to a closet wall and opened biometrically or with a key.

For defensive use, the proverbial "bump in the night" is a universal concern. That's no time to be fumbling with multiple locks or other complicated steps. Many people keep a handgun nearby. Regardless of your choice, a firearm is best located just far enough away from your bed to ensure full alertness upon access. There are a bunch of stories involving fathers shooting their teenage daughters at 2:00 AM. Give yourself time between a half-awake state to a loaded gun in your hand.

The accessories chapter discusses Remington's J-Lock safety button, which locks into the "on" position. Except for very small children, I'd skip it. The same applies to a trigger lock. If coupled with a locking case though, one would afford another layer of security when time is not a factor. This combination is often used during vehicular transport in more restrictive states. In such cases, lock everything out of sight, preferably in a separate cargo compartment.

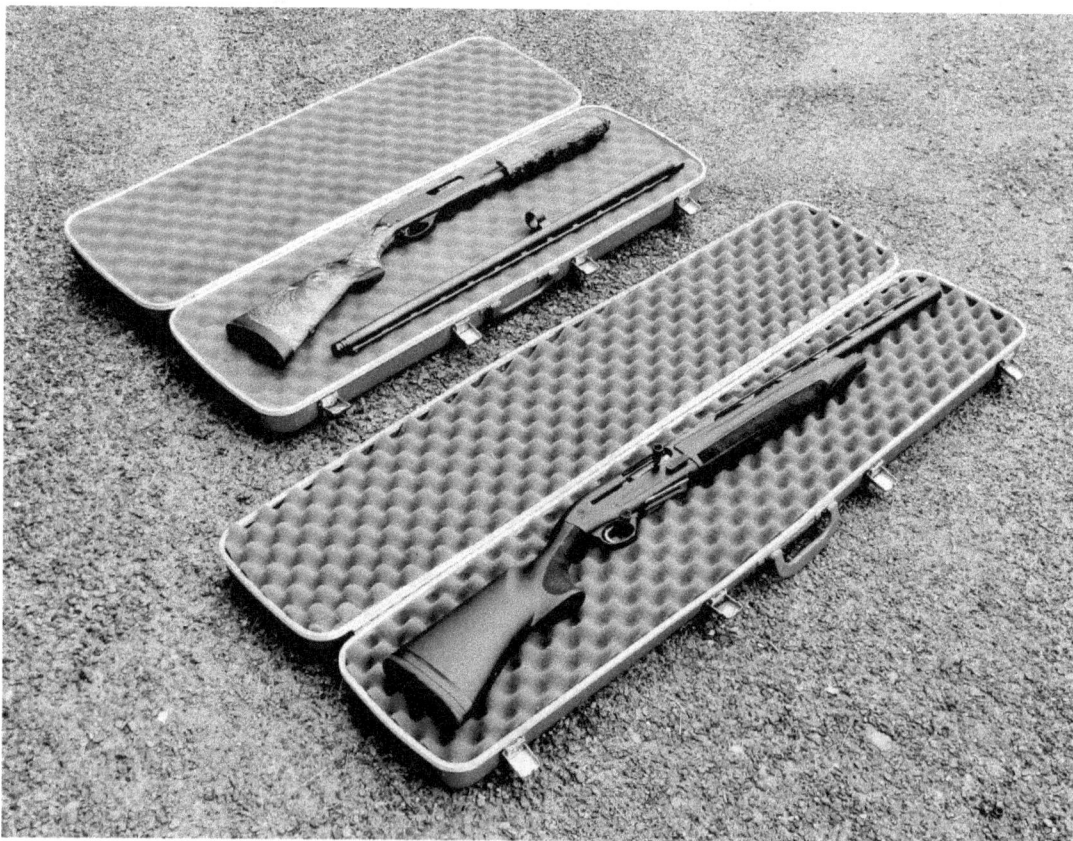

Locking hard-shell cases provide good protection for your guns but only marginal security. Stow them out of sight.

CARRY MODES

Whether afoot or mobile, muzzle control is always paramount! Unfortunately, some gun racks and tactical-type slings violate "rule two."

Vehicles. Somebody will probably wind up with a muzzle pointed at them during any horizontal carry. The racks in pickup truck windows come to mind, as do the forward mounts seen on ATV racks. We're okay with such positioning if an unloaded gun is inside a case, but that's about it. For those literally "riding shotgun" on a passenger side, muzzle up works.

A take-down case is designed for double-guns, but it will also work with many pumps or autos.

Preferably the gun will be unloaded, which is often a legal requirement. Another option is the carry condition load, involving an empty chamber. Either mode may be illegal in some states, so be sure to check. When driving solo, I prefer muzzle down stowage if quick access is needed. The gun goes on the passenger seat with its muzzle against the firewall. If a need to bail out arises, I can grab the gun by its grip with the barrel pointed away from me. Whatever you do, don't pull one toward you by the muzzle! Durable plastic scabbards, which are available for ATVs, locate a gun similarly while offering good protection from flying debris. They're a newer rendition of traditional horse-mounted versions.

Sooner or later someone will probably be swept by this muzzle.

A solo travel mode; note the open action.

Pickup trucks are the main means of transportation in my world, and a cased gun behind the seat is common. I wait to pull one out until any passengers are well clear of its business end. Leaning a gun against a vehicle is also ill-advised. Better to be safe than sorry, and a whole bunch of hunting accidents are related to sloppy handling.

Personal carry. All previous muzzle control concerns apply. Since a shotgun is not a hoe or a rake, please don't carry it like one. When in the company of others, you must be aware of your barrel at all times. Muzzle up or muzzle down carry is best, and port arms will suffice with care. An over-the-shoulder carry isn't a great idea in mixed company, nor is a crook-of-the-arm carry.

A sling is handy and I like strong-side, muzzle up carry during most instances. It'll work well with shorter barrels, but can be a nuisance with longer ones in thick woods. In that case, frequent dismounts will be needed to clear overhanging branches. Some people will grab the stock with their strong hand and pull the gun forward, swinging the muzzle end rearward in a pivoting arc. It's fast, but jeopardizes anyone located to the rear. Instead, we tell our troops to don or remove their gun like a coat, using the support-side hand to grasp the sling. That way the barrel remains pointing up.

When wearing a handgun on the strong-side hip, this carry can cause headaches. The handgun may catch on the sling, and its butt may even trip a slide release! Inclement weather is another concern since moisture will enter the muzzle. In fact, snow can be a problem if it dislodges from any overhanging branches. Some rifle hunters tape their muzzles, but the large opening of a 12-gauge makes it easier for the covering to enter the bore. An obstruction of any kind can be disastrous, sending shards of metal in all directions upon discharge.

Muzzle-up sling carry generally works well – but not with this gun in thick woods.

So, at times, opposite-side sling carry makes more sense. The muzzle goes down, the butt goes up, and the gun winds up behind the support-side shoulder. A fast dismount is possible by placing the support hand palm down on the forend while grasping it. The gun can then be pivoted up and forward with its stock trailing, under the arm. A big concern is the location of others in front of the shooter. Also, any muzzle-down carry can cause a serious bore obstruction through the introduction of snow. It's very easy to scoop some up in severe winter areas like ours.

The NRA will instruct you to unload when crossing any obstacles, which could include fences, stone walls, downed trees, or brooks. Tree stands should be negotiated only with a completely unloaded gun. The action should be open and the safety should be "on"!

Icy ground and rough terrain should be considered, too. They are good instances where carry condition is a safer bet. Travel through unsure footing increases the possibility of a fall and a hard impact can drop the hammer on a cocked gun. On the other hand, an empty chamber will remain that way if a gun like a Model 870 is cocked, which locks the slide forward. The safety should stay on as well. This

Muzzle-down, opposite-side carry works in the woods, but not in deep snow.

precaution may be a useful interim measure, balancing safety against the possible need for rapid use. Rack the slide, disengage the safety, and you're in action.

CUSTODY

As mentioned earlier, you're one-handed without a sling. Besides negotiating obstacles, two hands may be necessary at other times. With a sling, you can take the gun with you. (It could also serve as tourniquet during an emergency.) If you need to leave a gun behind, some measure of security is advised. Lacking any other means, secure it out of sight, unloaded, in a vehicle or other controlled space.

Disabling. One emergency technique that has been around a while involves stuffing the muzzle end into something capable of plugging the bore. In theory, any miscreant would recognize the gun as unsafe to fire. Not only that, but the ensuing results might take him out of the fight. This always seemed iffy to us, so we staged a controlled experiment, using a single shot, break-action shotgun slated for disposal. Bottom line: It *is* iffy! We plugged its muzzle with a few inches of mud and fired it remotely while it was aimed at a large target. The gun popped open, the muzzle split, and an egg-shaped swell developed behind it. Two 00 Buckshot holes appeared on the target, one of which was in its head. The distance was only five yards. Examining the gun indicated it might survive a second attempt. We repeated the test with similar results, which finally totaled the gun. Scratch this technique off the list.

However, without an extended magazine, it's easy to separate the barrel from the receiver of many repeating shotguns. You can even mash a barrel in a real bind. I once watched an officer pop out his trigger group using the shirt pin from his badge (it fit in his pocket)! They usually don't come out that easily, though. Running a set of handcuffs through an open ejection port prevents the bolt from closing and the other end can be secured to some immovable object.

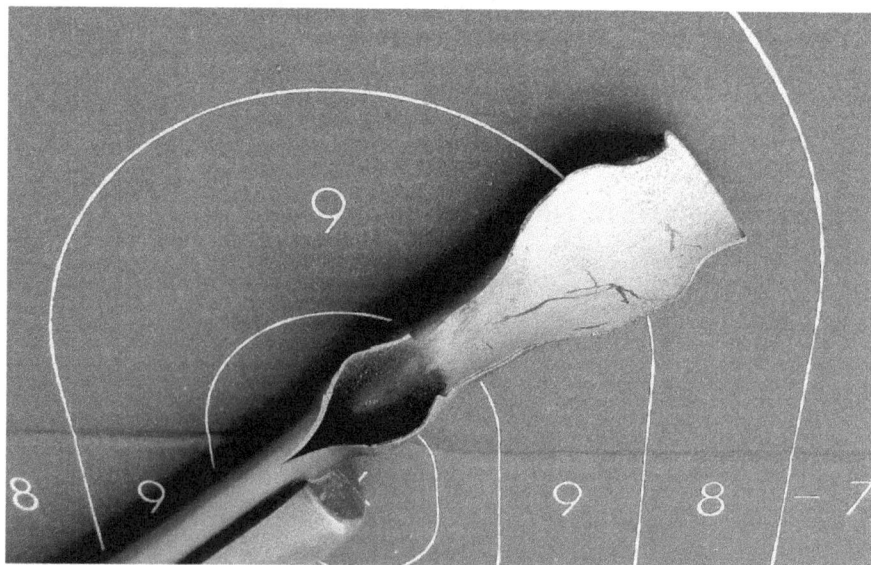

Test gun showing two rupture points. It still killed a paper man.

A cable lock, which may also deter theft if looped around an immoveable object.

A civilian equivalent is a hardened cable lock. Most of us won't have one in our pocket, but they'll work at pre-planned locations, including the home. For long-term use, some *unloaded* pump guns or autos can even be dry fired first. Carefully retract the bolt just far enough to feed through the cable without cocking the hammer. That way all of the primary springs will remain relaxed during long-term storage.

SUMMARY

Like any other firearm, a shotgun can cause serious damage. Firing just one load of shot into a clean target from inside five yards will result in a very large hole! Now picture that inside your home. Responsible ownership requires careful consideration of all potential liabilities. Please don't take any for granted.

Chapter 12

CONCLUSION

The finest sporting shotguns are elegant examples of gun makers' art, possessing magical qualities beyond description. The more utilitarian examples we've covered here share neither their grace, nor the mystique of the dreaded "black rifle." In fact, some are closer to appliances. But appliances do useful things and it's hard to ignore the versatility of a well-thought out smoothbore. From shot loads to slugs, within practical distances, the trusty scattergun will cover your six. For any doubters among us, rest assured that the sound of a slide running home is a universal language commanding immediate respect.

There's also just something intimidating about a .73 muzzle; especially if it moves toward your direction. Any emerging projectiles are cause for the gravest concerns and some will perform tasks not possible with any other weapon. Beyond conventional shot and slug loads, the a la carte menu includes specialty loads from less-lethal rounds to lock busters.

Furthermore, unlike the dreaded but erroneously labeled "assault rifles" demonized by both media and politicians, few legal restrictions apply to a garden-variety shotgun. Sure, a wood-stocked pump gun may be less glamorous, but it will also ring fewer alarm bells. Ironically, within reasonable distances, a shotgun is at least as formidable.

Beyond defense, it's hard to find a more universal tool for filling the family larder. Anything that flies, walks, or crawls is fair game with the right ammunition. Stock up on a few boxes of conventional shells and you'll be in business. Despite a chronic ammo shortage, 12 and 20-gauge ammunition remains readily available. Unlike many other types, cost has remained fairly stable. If things go bust, just about any hardware store will have a few boxes kicking around. For that matter, so might your uncle or grandpa.

The first gun in your safe really ought to be a shotgun. It will be among your most affordable choices and you won't go broke on accessories. Once equipped, you can save your pennies for a few more exotic toys. Meanwhile, you'll be suitably armed for nearly any contingency.

CHAPTER 13

OTHER FIREARMS MANUALS IN THE SURVIVAL GUNS SERIES

If you purchase a gun safe and attempt to fill it up in one fell swoop, you'll be hemorrhaging dollar bills. To keep things manageable, why not adopt an incremental approach? That's exactly what's been done with the succession of firearms manuals. You can focus on just one system and chip away until the essentials have been procured. The old saying, "a little knowledge is a dangerous thing" certainly holds true with firearms. Accordingly, each manual serves as a source for in-depth knowledge pertinent to specific systems. Furthermore, each is geared toward survival-based roles and the core principles espoused in *Survival Guns: A Beginner's Guide.*

SURVIVAL GUNS FIREARMS PUBLICATIONS IN PRINT

Survival Guns: A Beginner's Guide: This book is the first in the series, and serves as a guide to help build a basic firearm battery. It starts with a gun safe, to which firearms and accessories are added using a planned process. To help make the best choices, some key underlying principles are defined.

From there, procurement of several essential systems can commence. A baseline inventory of a shotgun, two rifles, and a handgun serve as cornerstones. Further additions include some interesting specialty firearms and accessories. The firearms on the essentials list, as well as many other types, will be thoroughly covered in the series of system-specific manuals. In each, the various models, ammunition, and accessories will be closely examined. While this book is written for beginners, those familiar with firearms should find topics of value. The information will be detailed, covering far more than just a firearm itself.

Air Rifles: A Buyer's and Shooter's Guide: Did you know that you can mail-order airguns in most locales? Unlike conventional firearms, they aren't federally regulated. They'll also get you into some tight places that would be strictly off limits to any powder-burning guns. The latest air-powered technologies are a quantum leap beyond a common BB gun, offering real quality and impressive performance. How about a rifle that runs on high-pressure air to combine effortless operation with multiple shots? These pre-charged types are filled from a scuba tank. Accuracy is phenomenal and so is power, yet noise is less than most silenced firearms. And again, no special BATF permits are necessary. Some can even be purchased as big-bore versions in .45 and .50 caliber. Others run independently from highly compressed springs or gas-strut type technologies. This publication provides the knowledge you'll need to fully exploit the advantages of these intriguing guns. A wide range of ammunition and powerplants are explored, along with their advantages and limitations. Scopes and other sighting systems are detailed, as are useful accessories. The airgun manual is your source for non-firearm technologies, from plinking and training through hunting.

Rimfire Rifles: A Buyer's and Shooter's Guide: This book is devoted to a whole series of useful firearms, beginning with the well-known .22 Long Rifle. The venerable "twenty-two" hosts a wide array of interesting loads, including some ultra-quiet choices and fairly nasty high-speed rounds. Even hotter rimfire calibers include the .22 Winchester Magnum Rimfire, plus three small-bore derivatives: the .17 Mach II, .17 Hornady Rimfire Magnum, and Winchester's high velocity .17 Super Magnum. Careful shopping can provide us with a useful rimfire (or maybe even two) with which to quietly harvest small game or eliminate pests. An economical .22 LR firearm can also serve as a great high-powered rifle trainer if similar function is considered. In a pinch, it might even work for self-defense. The rimfires can't do everything, but they can do a lot once fully understood. One trait is easy to overlook until you start shooting. They're lots of fun!

FUTURE SURVIVAL GUNS PUBLICATIONS

The Centerfire Rifle/AR-15 Manual: Are you interested in leverguns or semi-auto rifles? How about bolt-actions, or an entire family of guns in different calibers? How do you choose the right scope, and how do you sight it in? How do you select the most accurate loads, or pick the right bullets? How do you reduce recoil for younger shooters? Following in the steps of The Rimfire & Airgun Manual, the centerfire book takes us to the next level. Optics and ballistic aiming systems are explored, along with skill-building regimens. You'll see methods to assess true accuracy, and other useful tips. Different calibers and loads are discussed, as are various rifle choices. The AR-15 has grown wildly popular, with dozens of brands and hundreds of accessories to choose from. It's an extremely versatile platform for good reason, and can be instantly transformed to many different profiles. Switch-top conversions are possible in .22 LR, through several pistol calibers, and serious big-bore rounds. On top of that, universal "Picatinny" mounting points will easily accommodate nearly unlimited optical or equipment choices. Since most of us don't have unlimited funds, we'll need to determine exactly which accessories are necessary. This edition is the next progression in our system-based approach for development of a practical firearms battery.

The Handgun Manual: You can shoot yourself with the wrong combination of pistol, holster and clothing, so which ones are dangerous? You may understand the fundamentals of shooting, but how do they apply to handguns? Are you interested in a 1911 pistol? If so, did you know you can create your own multi-caliber pistol off a single frame? What about other types? Which loads are your best defensive choices? This publication covers everything you'll need to know about handguns, from different models through practical calibers, holsters, and accessories. You'll see some interesting alternatives to six-shot revolvers and the latest high-capacity pistols. The smaller guns are covered, too. Practical revolver and pistol skills are detailed, along with recommended practice regimens. The Handgun Manual rolls all of this information into one source for safe and effective handling.

INDEX

www.ingramcontent.com/pod-product-compliance
Lightning Source LLC
Chambersburg PA
CBHW080017280326
41934CB00015B/3372